WITHDRAWN

fracture mechanics

CIP-data Koninklijke Bibliotheek, Den Haag

Ewalds, H.L.

Fracture mechanics / H.L. Ewalds, R.J.H. Wanhill. – Delft: Delftse U.M.; Londen [etc.]:
Arnold. – Ill., graf. With ref., index
ISBN 90-6562-024-9 (DUM)
ISBN 0-7131-3515-8 (Arnold)
SISO 533 UDC 531
Subject heading: fracture mechanics.

© **VSSD**

Third imprint 1986. First published in 1984 by
Edward Arnold (Publishers) Ltd
41 Bedford Square, London WC1B 3DQ UK
3 East Read Street, Baltimore, Maryland 21202 USA

Edward Arnold (Australia) Pty Ltd
80 Waverley Road, Caulfield East, Victoria 3145 Australia

and

Delftse Uitgevers Maatschappij b.v.
P.O. Box 2851, 2601 CW Delft, The Netherlands

Arnold edition: ISBN 0 7131 3515 8
DUM edition: ISBN 90 6562 024 9

fracture mechanics

H.L. Ewalds
Boiler and Pressure Vessel Authority
'Dienst voor het Stoomwezen', The Hague

R.J.H. Wanhill
National Aerospace Laboratory NLR,
Emmeloord and Amsterdam

co-publication of

 Edward Arnold    Delftse Uitgevers Maatschappij

Fixed Grips and Constant Load

Often we must show — to the best of our ability —
That nothing can result in crack growth instability.
Precision of analysis is unhappily elastic,
Owing to the problem that most strains are really plastic.
An added burden to each crack's calamitous position
Is doubt about fixed grips and the constant load condition.
U *plus* F *is* U_0 *plus* U_a *plus* U_γ
This equation is so simple, we speak it without stammer.
Now change the crack length slightly, increment da,
At once we're not so certain of what we have to say.
The energy release rate must always equal G,
But positive or negative? It's difficult to see.
So when you read our text on this in section four point two
We hope there's not much trace of all that we have suffered through!

R.J.H. Wanhill

Preface

While teaching a course on fracture mechanics at Delft University of Technology we discovered that although there are a few excellent textbooks, their subject matter covers developments only up to the early 1970s. Consequently there was no systematic treatment of the concepts of elastic-plastic fracture mechanics. Also the description of fracture mechanics characterization of crack growth needed updating, especially for sustained load fracture and unstable dynamic crack growth.

In the present textbook we have attempted to cover the basic concepts of fracture mechanics for both the linear elastic and elastic-plastic regimes, and three chapters are devoted to the fracture mechanics characterization of crack growth (fatigue crack growth, sustained load fracture and dynamic crack growth).

There are also two chapters concerning mechanisms of fracture and the ways in which actual material behaviour influences the fracture mechanics characterization of crack growth. The reader will find that this last topic is treated to some way beyond that of a basic course. This is because to our knowledge there is no reference work that systematically covers it. A consequence for instructors is that they must be selective here. However, any inconvenience thereby entailed is, we feel, outweighed by the importance of the subject matter.

This textbook is intended primarily for engineering students. We hope it will be useful to practising engineers as well, since it provides the background to several new design methods, criteria for material selection and guidelines for acceptance of weld defects.

Many people helped us during preparation of the manuscript. We wish to thank particularly J. Zuidema, who made vital contributions to uniform treatment of the energy balance approach for both the linear elastic and elastic-plastic regimes; R.A.H. Edwards, who assisted with the chapter on sustained load fracture; A.C.F. Hagedorn, who drew the figures for the first seven chapters; and the team of the VSSD, our publisher, whose patience was sorely tried but who remained unbelievably cooperative.

Finally, we wish to thank the National Aerospace Laboratory NLR and the Boiler and Pressure Vessel Authority 'Dienst voor het Stoomwezen' for providing us the opportunity to finish this book, which was begun at the Delft University of Technology.

The Authors
September 1983

Contents

Part I Introduction

1. AN OVERVIEW

1.1. About this Course

This course is intended as a basic background in fracture mechanics for engineering usage. In order to compile the course we have consulted several textbooks and numerous research articles. In particular, the following books have been most informative and are recommended for additional reading:

- D. Broek, "Elementary Engineering Fracture Mechanics", Martinus Nijhoff (1982) The Hague.
- J.F. Knott, "Fundamentals of Fracture Mechanics", Butterworths, (1973) London.
- "A General Introduction to Fracture Mechanics", Institution of Mechanical Engineers (1978) London.

As indicated in the table of contents the course has been divided into several parts. Part I, consisting of this chapter, is introductory. In Part II the well established subject of Linear Elastic Fracture Mechanics (LEFM) is treated, and this is followed in Part III by the more recent and still evolving topic of Elastic-Plastic Fracture Mechanics (EPFM). In Part IV the applicability of fracture mechanics concepts to crack growth behaviour is discussed: namely subcritical, stable crack growth under cyclic loading (fatigue) or sustained load, and dynamic crack growth beyond instability. Finally, in Part V the mechanisms of fracture in actual materials are described together with the influence of material behaviour on fracture mechanics-related properties.

1.2. Historical Review

Strength failures of load bearing structures can be either of the yielding-dominant or fracture-dominant types. Defects are important for both types of failure, but those of primary importance to fracture differ in an extreme way from those influencing yielding and the resistance to plastic flow. These differences are illustrated schematically in figure 1.1.

For yielding-dominant failures the significant defects are those which tend to warp and interrupt the crystal lattice planes, thus interfering with dislocation glide and providing a resistance to plastic deformation that is essential to the strength of high strength metals. Examples of such defects are interstitial and out-of-size substitutional atoms, grain boundaries, coherent precipitates and dislocation networks. Larger defects like inclusions, porosity, surface scratches and small cracks may influence the effective net section bearing the load, but otherwise have little effect on resistance to yielding.

For fracture-dominant failures, i.e. fracture before general yielding of the net section, the size scale of the defects which are of major significance is essentially macroscopic, since general plasticity is not involved but only the local stress-strain fields associated with the defects. The minute lattice-related defects which control resistance to plastic flow are not of direct concern.

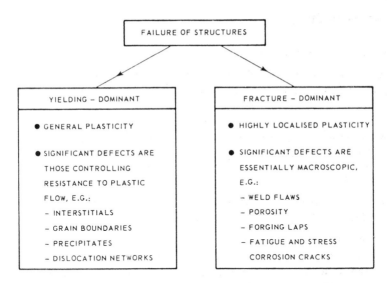

Figure 1.1. Types of structural failure.

They are important insofar as the resistance to plastic flow is related to the material's susceptibility to fracture.

Fracture mechanics, which is the subject of this course, is concerned almost entirely with fracture-dominant failure. The first successful analysis of a fracture-dominant problem was that of Griffith in 1920, who considered the propagation of brittle cracks in glass. Griffith formulated the now well-known concept that an existing crack will propagate if thereby the total energy of the system is lowered, and he assumed that there is a simple energy balance, consisting of a decrease in elastic strain energy within the stressed body as the crack extends, counteracted by the energy needed to create the new crack surfaces. His theory allows the estimation of the theoretical strength of brittle solids and also gives the correct relationship between fracture strength and defect size.

The Griffith concept was first related to brittle fracture of metallic materials by Zener and Hollomon in 1944. Soon after, Irwin pointed out that the Griffith-type energy balance must be between (1) the stored strain energy and (2) the surface energy plus the work done in plastic deformation. Irwin also recognized that for relatively ductile materials the energy required to form new crack surfaces is generally insignificant compared to the work done in plastic deformation, and he defined a material property G as the total energy absorbed during cracking per unit increase in crack length and per unit thickness. G is called the 'energy release rate' or 'crack driving force'.

In the middle 1950s Irwin contributed another major advance by showing that the energy approach is equivalent to a stress intensity (K) approach, according to which fracture occurs when a critical stress distribution ahead of the crack tip is reached. The material property governing fracture may therefore be stated as a critical stress intensity, K_c, or in terms of energy as a critical value G_c.

Demonstration of the equivalence of G and K provided the basis for development of the discipline of Linear Elastic Fracture Mechanics (LEFM). This is because the form of the stress distribution around and close to a crack tip is always the same. Thus tests on suitably shaped and loaded specimens to determine K_c make it possible to determine what flaws are tolerable in an actual structure under given conditions. Furthermore, materials can be compared as to their utility in situations where fracture is possible. It has also been found that the sensitivity of structures to subcritical cracking such as fatigue crack growth and stress corrosion can, to some extent, be predicted on the basis of tests using the stress intensity approach.

The beginnings of Elastic—Plastic Fracture Mechanics (EPFM) can be traced to fairly early in the development of LEFM, notably Wells' work on Crack Opening Displacement (COD), which was published in 1961. However, the greater complexity of the problems of analysis has necessarily led to somewhat slower progress. EPFM is still very much an evolving discipline.

1.3. The Significance of Fracture Mechanics

In the nineteenth century the Industrial Revolution resulted in an enormous increase in the use of metals (mainly irons and steels) for structural applications. Unfortunately, there also occurred many accidents, with loss of life, owing to failure of these structures. In particular, there were numerous accidents involving steam boiler explosions and railway equipment.

Some of these accidents were due to poor design, but it was also gradually discovered that material deficiencies in the form of pre-existing flaws could initiate cracking and fracture. Prevention of such flaws by better production methods reduced the number of failures to more acceptable levels.

A new era of accident-prone structures was ushered in by the advent of all-welded designs, notably the Liberty ships and T-2 tankers of World War II. Out of 2500 Liberty ships built during the war, 145 broke in two and almost 700 experienced serious failures. Many bridges and other structures also failed. The failures often occurred under very low stresses, for example even when

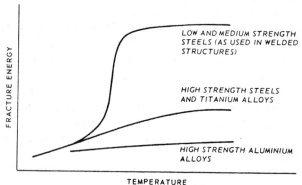

Figure 1.2. Schematic of the general effect of temperature on the fracture resistance of structural metals.

a ship was docked, and this anomaly led to extensive investigations which revealed that the fractures were brittle and that flaws and stress concentrations were responsible. It was also discovered that brittle fracture in the types of steel used was promoted by low temperatures. This is depicted in figure 1.2: above a certain transition temperature the steels behave in a ductile manner and the energy required for fracture increases greatly.

Current manufacturing and design procedures can prevent the intrinsically brittle fracture of welded steel structures by ensuring that the material has a suitably low transition temperature and that the welding process does not raise it. Nevertheless, service-induced embrittlement, for example irradiation effects in nuclear pressure vessels and corrosion fatigue in offshore platforms, remains a cause for concern.

Looking at the present situation it may be seen from figure 1.3 that since World War II the use of high strength materials for structural applications has greatly increased.

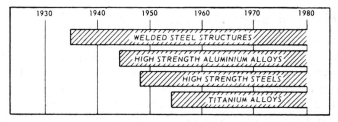

Figure 1.3. Introduction of high strength materials for structural applications.

These materials are often selected to obtain weight savings — aircraft structures are an obvious example. Additional weight savings have come from refinements in stress analysis, which have enabled design allowables to be raised. However, it was not recognized until towards the end of the 1950s that although these materials are not intrinsically brittle, the energy required for fracture is comparatively low, as figure 1.2 shows. The possibility, and indeed occurrence, of this low energy fracture in high strength materials stimulated the modern development of fracture mechanics.

The object of fracture mechanics is to provide quantitative answers to specific problems concerning cracks in structures. As an illustration, consider a structure containing pre-existing flaws and/or in which cracks initiate in service. The cracks may grow with time owing to various causes (for example fatigue, stress corrosion, creep) and will generally grow progressively faster, figure 1.4.a. The residual strength of the structure, which is the failure strength as a function of crack size, decreases with increasing crack size, as shown in figure 1.4.b. After a time the residual strength becomes so low that the structure may fail in service.

With respect to figure 1.4 fracture mechanics should attempt to provide quantitative answers to the following questions:

1) What is the residual strength as a function of crack size?
2) What crack size can be tolerated under service loading, i.e. what is the

16

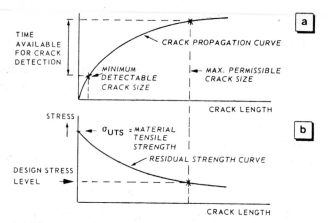

Figure 1.4. The engineering problem of a crack in a structure.

maximum permissible crack size?

3) How long does it take for a crack to grow from a certain initial size, for example the minimum detectable crack size, to the maximum permissible crack size?

4) What is the service life of a structure when a certain pre-existing flaw size (e.g. a manufacturing defect) is assumed to exist?

5) During the period available for crack detection how often should the structure be inspected for cracks?

This course is intended to show how fracture mechanics concepts can be applied so that these questions can be answered.

In the remaining sections 1.4 − 1.10 of this introductory chapter an overview of basic concepts and applications of LEFM are given in preparation for more detailed treatment in subsequent chapters.

1.4. The Griffith Energy Balance Approach

Consider an infinite plate of unit thickness that contains a through-thickness crack of length 2a and that is subjected to uniform tensile stress, σ, applied at infinity. Figure 1.5 represents an approximation to such a plate.

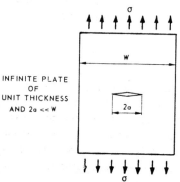

Figure 1.5. A through-cracked plate.

The total energy U of the cracked plate may be written as

$$U = U_0 + U_a + U_\gamma - F, \qquad (1.1)$$

where U_0 = elastic energy of the loaded uncracked plate (a constant),

U_a = change in the elastic energy caused by introducing the crack in the plate,

U_γ = change in the elastic surface energy caused by the formation of the crack surfaces,

F = work performed by external forces: this must be substracted in equation (1.1), since it is not part of the internal (potential) energy of the plate. $F = \text{load} \times \text{displacement}$.

Griffith used a stress analyses developed by Inglis to show that for unit thickness the absolute value of U_a is given by

$$|U_a| = \frac{\pi \sigma^2 a^2}{E}. \qquad (1.2)$$

Moreover, the elastic surface energy, U_γ, is equal to the product of the elastic surface energy of the material, γ_e, and the new surface area of the crack:

$$U_\gamma = 2(2a\gamma_e). \qquad (1.3)$$

For the case where no work is done by external forces, the so-called fixed grip condition, $F = 0$ and the change in elastic energy U_a, caused by introducing the crack in the plate, is negative: there is a decrease in elastic strain energy of the plate because it loses stiffness and the load applied by the fixed grips will therefore drop. Consequently, the total energy U of the cracked plate is

$$U = U_0 + U_a + U_\gamma$$

$$= U_0 - \frac{\pi \sigma^2 a^2}{E} + 4a\gamma_e. \qquad (1.4)$$

Since U_0 is constant, dU_0/da is zero, and the equilibrium condition for crack extension is obtained by setting dU/da equal to zero

$$\frac{d}{da}(-\frac{\pi \sigma^2 a^2}{E} + 4a\gamma_e) = 0. \qquad (1.5)$$

This is illustrated in figure 1.6. Figure 1.6.a schematically represents the two energy terms in equation (1.5) and their sum as functions of the introduced crack length, 2a. Figure 1.6.b represents the derivative, dU/da. When the elastic energy release due to a potential increment of crack growth, da, outweighs the demand for surface energy for the same crack growth, the introduction of a crack will lead to its unstable propagation.

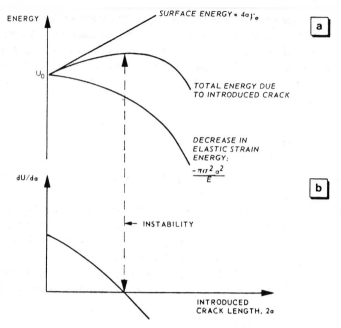

Figure 1.6. Energy balance of a crack in an infinite plate when no work is done by external forces.

From the equilibrium condition one obtains

$$\frac{2\pi\sigma^2 a}{E} = 4\gamma_e.$$ (1.6)

which can be rearranged to

$$\sigma\sqrt{a} = (\frac{2E\gamma_e}{\pi})^{\frac{1}{2}}.$$ (1.7)

Equation (1.7) indicates that crack extension in ideally brittle materials is governed by the product of the remotely applied stress and the square root of the crack length and by material properties. Because E and γ_e are material properties the right-hand side of equation (1.7) is equal to a constant value characteristic of a given ideally brittle material. Consequently, equation (1.7) indicates that crack extension in such materials occurs when the product $\sigma\sqrt{a}$ attains a constant critical value.

1.5. Irwin's Modification to the Griffith Theory

Equation (1.6) may be rearranged in the form

$$\frac{\pi\sigma^2 a}{E} = 2\gamma_e.$$ (1.8)

The left-hand side of equation (1.8) has been designated the energy release rate, G, and represents the elastic energy per unit crack surface area that is available for infinitesimal crack extension. The right-hand side of equation (1.8) represents the surface energy increase that would occur owing to infinitesimal crack extension and is designated the crack resistance, R. It follows that G must be at least equal to R before unstable crack growth occurs. If R is a constant, this means that G must exceed a critical value G_c. Thus fracture occurs when

$$\frac{\pi \sigma^2 a}{E} \geqslant \frac{\pi \sigma_c^2 a}{E} = G_c = R. \tag{1.9}$$

The critical value G_c can be determined by measuring the stress σ_c required to fracture a plate with a crack of size 2a.

In 1948 Irwin suggested that the Griffith theory for ideally brittle materials could be modified and applied to both brittle materials and metals that exhibit plastic deformation. A similar modification was proposed by Orowan. The modification recognized that a material's resistance to crack extension is equal to the sum of the elastic surface energy and the plastic strain work, γ_p, accompanying crack extension. Consequently, equation (1.8) was modified to

$$\frac{\pi \sigma^2 a}{E} = 2(\gamma_e + \gamma_p). \tag{1.10}$$

For relatively ductile materials $\gamma_p \gg \gamma_e$, i.e. R is mainly plastic energy and the surface energy can be neglected.

Although Irwin's modification includes a plastic energy term, the energy balance approach to crack extension is still limited to defining the conditions required for instability of an ideally sharp crack. Also, the energy balance approach presents insuperable problems for many practical situations, especially slow stable crack growth, as for exemple in fatigue and stress corrosion cracking.

The energy balance concept will be treated in more detail in chapter 4.

1.6. The Stress Intensity Approach

Owing to the practical difficulties of the energy approach a major advance was made by Irwin in the 1950s when he developed the stress intensity approach. First, from linear elastic theory Irwin showed that the stresses in the vicinity of a crack tip take the form

$$\sigma_{ij} = \frac{K}{\sqrt{2\pi r}} \, f_{ij}(\theta) + \ldots \tag{1.11}$$

where r, θ are the cylindrical polar coordinates of a point with respect to the crack tip, figure 1.7.

K is a constant which gives the magnitude of the elastic stress field. It is called the *stress intensity factor*. Dimensional analysis shows that K must be linearly related to stress and direcly related to the square root of a characteristic length. Equation (1.7) from Griffith's analysis indicates that this characteristic length is the crack length, and it turns out that the general form of the stress

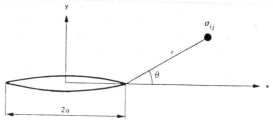

Figure 1.7. Stresses at a point ahead of a crack tip.

intensity factor is given by

$$K = \sigma \sqrt{\pi a} \cdot f(\frac{a}{W})$$ (1.12)

where $f(\frac{a}{W})$ is a dimensionless parameter that depends on the geometries of the specimen and crack.

Irwin then demonstrated that if a crack is extended by an amount da, the work done by the stress field ahead of the crack when moving through the displacements corresponding to a crack of length (a + da) is formally equivalent to the change in strain energy G da. Thus the achievement of a critical stress intensity factor, K_c, is exactly equivalent to the Griffith-Irwin energy balance approach, which requires the achievement of a stored elastic strain energy equal to G_c.

The parameter governing fracture may therefore be stated as a critical stress intensity, K_c, instead of a critical energy value G_c. For tensile loading the relationships between K_c and G_c are

$$G_c = \frac{K_c^2}{E} \quad \text{plane stress;}$$

$$G_c = \frac{K_c^2}{E} (1 - \nu^2) \quad \text{plane strain.}$$ (1.13)

where ν is Poisson's ratio. For plane strain it is customary to write G_{Ic} and K_{Ic}, where the subscript I indicates tensile loading. This subscript is also used for expressions including the stress intensity factor as a variable, i.e. K_I.

The Griffith-Irwin solution for a through-cracked plate can now be written as

$$\sigma \sqrt{\pi a} = (2E(\gamma_e + \gamma_p))^{\frac{1}{2}} = (EG)^{\frac{1}{2}} = K$$ (1.14)

and so the failure criterion is

$$\sigma \sqrt{\pi a} \geqslant \sigma_c \sqrt{\pi a} = K_c.$$ (1.15)

More generally, however, it is to be noted that as in the Griffith approach the use of a critical stress intensity indicates that crack extension occurs when the product $\sigma \sqrt{a}$ attains a constant critical value. The value of this constant can be determined experimentally by measuring the fracture stress for a large plate that contains a through-thickness crack of known length. This value can also be measured by using other specimen geometries, or else can be used to predict critical combinations of stress and crack length in these other geometries. This is what makes the stress intensity approach to fracture so powerful, since values of K for different specimen geometries can be determined from conventional elastic stress analyses: there are now several handbooks giving relationships between the stress intensity factor and many types of cracked bodies with different crack sizes, orientations and shapes, and loading conditions. Furthermore, the stress intensity factor, K, is applicable to stable crack extension and does to some extend characterize processes of subcritical cracking like fatigue and stress corrosion, as will be mentioned in section 1.9 of this chapter and in greater detail in chapters 9 and 10.

It is the use of the stress intensity factor as the characterizing parameter for crack extension that is the fundamental principle of Linear Elastic Fracture Mechanics (LEFM). The theory of Linear Elastic Fracture Mechanics is well developed and will be discussed in chapter 2.

1.7. Crack Tip Plasticity

The elastic stress distribution in the vicinity of a crack tip, equation (1.11), shows that as r tends to zero the stresses become infinite, i.e. there is a stress singularity at the crack tip. Since structural materials deform plastically above the yield stress, there will in reality be a plastic zone surrounding the crack tip. Thus the elastic solution is not unconditionally applicable.

Irwin considered a circular plastic zone to exist at the crack tip under tensile loading. As will be discussed in chapter 3, he showed that such a circular plastic zone has a diameter $2r_y$, figure 1.8.a, with

$$r_y = \frac{1}{2\pi} \left(\frac{K_I}{\sigma_{ys}}\right)^2 \quad \text{plane stress} \tag{1.16}$$

and

$$r_y = \frac{1}{2\pi} \left(\frac{K_I}{C\sigma_{ys}}\right)^2 \quad \text{plane strain} \tag{1.17}$$

where C is usually estimated to be about 1.7.

Irwin argued that the occurrence of plasticity makes the crack behave as if it were longer than its physical size — the displacements are larger and the stiffness is lower than in the elastic case. He showed that the crack may be viewed as having a notional tip a distance r_y ahead of the real tip, figure 1.8.b, with a yielded region r_y beyond this and the K distribution of local stress taking over from the yield stress at a distance r_y ahead of the notional crack tip. Since the same K always gives the same plastic zone size for materials with the same yield stress, equations (1.16) and (1.17), the stresses and strains both within and outside the plastic zone will be determined by K and the stress intensity approach can still be used. In short, the effect of crack tip

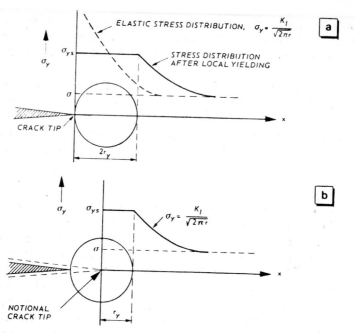

Figure 1.8. The crack tip plastic zone and Irwin's correction for it.

plasticity corresponds to an apparent increase of the elastic crack length by an increment equal to r_y.

It must be noted that the scope of Linear Elastic Fracture Mechanics can be extended to cope with only limited crack tip plasticity, i.e. when the plastic zone is small compared to the crack size and the cracked body still behaves in an approximately elastic manner. If this is not the case then the problem has to be treated elasto-plastically. However, the concepts of Elastic–Plastic Fracture Mechanics (EPFM) are not nearly so well developed as LEFM theory, and this fact is reflected in the approximate nature of the eventual solutions.

The two most promising concepts of EPFM are the Crack Opening Displacement (COD) approach, based on crack tip strain, and the J integral energy balance concept. These will be described in chapters 6–8.

1.8. Fracture Toughness

From sections 1.6 and 1.7 it follows that under conditions of limited crack tip plasticity the parameter governing tensile fracture can be stated as a critical stress intensity, either K_c (plane stress) or K_{Ic} (plane strain). The value of K_c at a particular temperature depends on specimen thickness and constraint. The limiting value of K_c for maximum constraint (plane strain) is K_{Ic}.

K_{Ic} can be considered a material property characterizing the crack resistance, and is therefore called the plane strain fracture toughness. Thus the same value of K_{Ic} should be found by testing specimens of the same material with different geometries and with critical combinations of crack size and shape and fracture stress. Within certain limits this is indeed the case, and so a knowledge of K_{Ic} obtained under standard conditions can be used to predict failure for different combinations of stress and flaw size and for different geometries.

K_c can also be determined under standard conditions, and the value thus found may also be used to predict failure, but only for situations with the same material thickness and constraint.

As an introductory numerical example of the design application of LEFM, consider the equation for a through-thickness crack in a wide plate, i.e.

$$K = \sigma \sqrt{\pi a}. \tag{1.18}$$

Assume that the test results show that for a particular steel the K_c is 66 MPa$\sqrt{m}$ for the plate thickness and temperature in service. Using equation (1.18) a residual strength curve for this steel can be constructed relating K_c and nominal stress and crack size. This is shown in figure 1.9. Also assume that the design stress is 138 MPa. It follows from equation (1.18) and figure 1.9 that the tolerable flaw size would be about 145 mm. For a design stress of 310 MPa the same material could tolerate a flaw size of only about 28 mm. Note from figure 1.9 that if a steel with a higher fracture toughness is used, for example one with a K_c of 132 MPa$\sqrt{m}$, the permissible design stress for a given flaw size is significantly increased, i.e. the material with the higher fracture toughness has the higher residual strength.

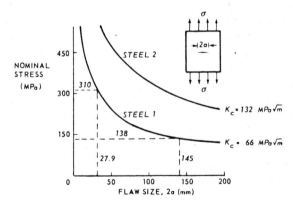

Figure 1.9. Residual strength curves for two steels.

1.9. Subcritical Crack Growth

In section 1.7 it was mentioned that the stress intensity factor can still be used when crack tip plasticity is limited. This latter condition holds for subcritical crack growth, where most of the crack extension usually takes place at stress intensities well below K_c and K_{Ic}, and it has been found that the stress intensity approach can provide correlations of data for fatigue crack growth and stress corrosion cracking.

Fatigue

Consider a through-thickness crack in a wide plate subjected to remote stressing that varies cyclically between constant minimum and maximum values, i.e. a fatigue loading consisting of constant amplitude stress cycles as in figure 1.10. The stress range $\Delta\sigma = \sigma_{max} - \sigma_{min}$, and from equation (1.18) a stress intensity factor range may be defined:

$$\Delta K = K_{max} - K_{min} = \Delta\sigma\sqrt{\pi a} \qquad (1.19)$$

where the subscript I to indicate tensile loading is customarily omitted.

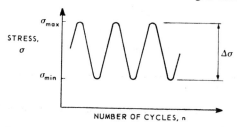

Figure 1.10. Stress-cycle parameters in constant amplitude fatigue.

The fatigue crack propagation rate is defined as the crack extension, Δa, during a small number of cycles, Δn, i.e. the propagation rate is $\Delta a/\Delta n$, which in the limit can be written as the differential da/dn. It has been found experimentally that provided the stress ratio, $R = \sigma_{min}/\sigma_{max}$, is the same then ΔK correlates fatigue crack growth rates in specimens with different stress ranges and crack lengths and also correlates crack growth rates in specimens of different geometry, i.e.

$$\frac{da}{dn} = f(\Delta K, R). \qquad (1.20)$$

This correlation is shown schematically in figure 1.11. Note that it is customary to plot $da/dn - \Delta K$ data on a double logarithmic diagram. The data obtained with a high stress range, $\Delta\sigma_h$, commence at relatively high values of da/dn and ΔK. The data for a low stress range, $\Delta\sigma_l$, commence at lower values of da/dn and ΔK, but reach the same high values as in the high stress range case. The data frequently show a sigmoidal trend, and this will be discussed in chapter 9 together with additional aspects of fatigue crack growth.

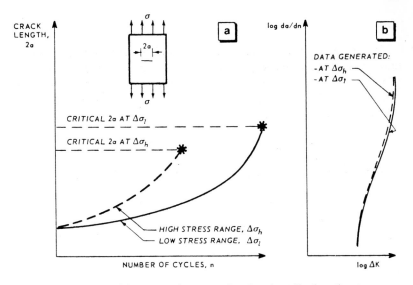

Figure 1.11. Correlation of fatigue crack propagation data by ΔK when the stress ratio, R, is the same.

Stress Corrosion

It has also been found that stress corrosion cracking data may be correlated by the stress intensity approach. Figure 1.12 gives a generalized representation of the stress corrosion crack growth rate, da/dt, as a function of K_I, where t is time.

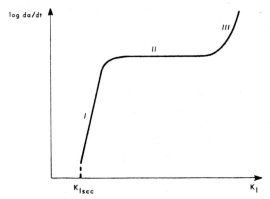

Figure 1.12. Stress corrosion crack growth rate as a function of K_I.

The crack growth curve consists of three regions. In regions I and III the crack velocity depends strongly on K_I, but in region II the velocity is virtually independent of K_I. Regions I and II are most characteristic. Region III is often not observed owing to a fairly abrupt transition from region II to unstable fast fracture. In region I there is a so-called threshold stress intensity, designated K_{Iscc}, below which cracks do not propagate under sustained load for a given combination of material and environment. This threshold stress intensity is an important parameter that can be determined by time-to-failure tests in which pre-cracked specimens are loaded at various stress intensity levels, thereby failing at different times as shown schematically in figure 1.13.

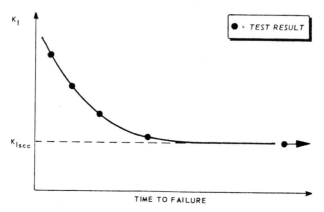

Figure 1.13. Schematic time-to-failure curve with K_{Iscc}.

The subject of stress corrosion cracking, under the more general heading of sustained load fracture, will be examined further in chapter 10.

1.10. Influence of Material Behaviour

So far, this overview of the use of fracture mechanics to characterize crack extension has not taken account of actual material behaviour, the influence of which may be considerable. For example, the fracture toughness of a material is much less when crack extension occurs by cleavage instead of ductile fracture. Cleavage is an intrinsically brittle mode of fracture involving separation of atomic bonds along well-defined crystallographic planes.

Other examples of material behaviour that affects fracture properties are:
1) Cracking of second phase particles in the metallic matrix and formation of microvoids at particle/matrix interfaces.
2) Anisotropic deformation and fracture. This may be intrinsic (crystallographic) as in the case of cleavage, or may result from material processing.
3) Choice of fracture path, i.e. whether transgranular or intergranular, or a mixture of both.
4) Crack blunting and branching.

In fact, fracture often depends on combinations of such types of material behaviour. For this reason we consider that a basic course in fracture mechanics should include information concerning mechanisms of fracture and the influence of material behaviour on fracture mechanics-related properties. These topics are dealt with in chapters 12 and 13.

Part II Linear Elastic Fracture Mechanics

2. THE ELASTIC STRESS FIELD APPROACH

2.1. Introduction

In the overview given in chapter 1 it was stated that the stress intensity factor K describes the magnitude of the elastic crack tip stress field. Also, K can correlate the crack growth and fracture behaviour of materials provided that the crack tip stress field remains predominantly elastic. This correlating ability makes the stress intensity factor an extremely important fracture mechanics parameter. For this reason its derivation is treated in some detail in section 2.2.

All stress systems in the vicinity of a crack tip may be derived from three modes of loading, figure 2.1. In what follows the derivation of elastic stress field equations will be limited to mode I, since this is the predominant stress situation in many practical cases. Once this derivation is understood it is possible to obtain a number of useful expressions for stresses and displacements in the crack tip region. However, use of the stress intensity factor approach for practical geometries does involve some difficulties. For example, actual cracks may be very irregular in shape as compared to the often highly idealised cracks considered in theoretical treatments. Moreover, assumptions such as the infinite width of a sheet or plate frequently cannot be maintained if an accurate result is required. The consequences of necessary deviations from the theoretical solutions will also be discussed in this chapter.

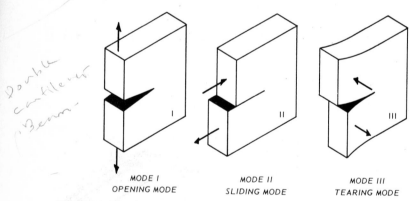

Double
cantilever
Beam.

| MODE I | MODE II | MODE III |
| OPENING MODE | SLIDING MODE | TEARING MODE |

Figure 2.1. The three modes of loading.

2.2. Derivation of the Mode I Elastic Stress Field Equations

This section gives an overview of the derivation of the mode I stress field equations. More rigorous treatments may be found in references 1 and 2 of the bibliography at the end of this chapter. The derivation covers the following topics:

- The equilibrium equations of stresses and compatibility equation of strains
- The Airy stress functions
- The Westergaard solution
- Complex variables and the Cauchy-Riemann equations
- Boundary conditions
- Use of polar coordinates
- Final solution for the stress intensity factor.

After the solution for the stress intensity factor has been obtained some additional remarks concerning the derivation are made.

To solve any plane elastic problem the equilibrium equations of stresses and the compatibility equations of strain must be obeyed.

Airy demonstrated that for any elastic problem there exists a function $\Phi(x,y)$ which fulfils the equilibrium equations of stresses. The stresses σ_x, σ_y and τ_{xy} are defined as

$$\sigma_x = \frac{\partial^2 \Phi}{\partial y^2}$$

$$\sigma_y = \frac{\partial^2 \Phi}{\partial x^2} \tag{2.1}$$

$$\tau_{xy} = -\frac{\partial^2 \Phi}{\partial x \partial y} \; .$$

For the compatibility equation of strains the following biharmonic equation can then be derived

$$\nabla^4(\Phi) = \nabla^2 [\nabla^2(\Phi)] = 0. \tag{2.2}$$

Every function $\Phi(x,y)$ that fulfils equation (2.2) is called an (Airy) stress function.

Consider now the specific problem of an infinite, biaxially loaded plate containing a crack. To solve this problem we can take a complex stress function of a type introduced by Westergaard:

$$\Phi = \text{Re}\, \bar{\bar{\phi}}(z) + y\, \text{Im}\, \bar{\phi}(z). \tag{2.3}$$

(In equation (2.3) $\phi(z)$ is an analytic function of the complex variable z $(x+iy)$, and $\bar{\bar{\phi}}(z)$ and $\bar{\phi}(z)$ are the first and second order integrals. It can be proved that the real and imaginary parts of such an analytic function fulfil the biharmonic equation, and that the products of these parts with the variables x and y also do so. Thus analytic functions like that in equation (2.3) may be used as stress functions.)

Using the Cauchy-Riemann equations

$$\frac{\partial(\text{Re } f(z))}{\partial x} = \frac{\partial(\text{Im } f(z))}{\partial y}$$

and (2.4)

$$\frac{\partial(\text{Re } f(z))}{\partial y} = -\frac{\partial(\text{Im } f(z))}{\partial x}$$

it is possible to find expressions for σ_x, σ_y and τ_{xy} by differentiating equation (2.3) according to the rules of equation (2.1). This gives

$$\sigma_x = \text{Re } \phi(z) - y \text{ Im } \phi'(z)$$

$$\sigma_y = \text{Re } \phi(z) + y \text{ Im } \phi'(z) \qquad (2.5)$$

$$\tau_{xy} = -y \text{ Re } \phi'(z)$$

where $\phi'(z)$ is the first order derivative. Note that equations (2.5) are general solutions which will give stresses for any $\phi(z)$. However, the correct stresses for a particular problem will be obtained only by using a function $\phi(z)$ that fulfils a number of boundary conditions pertaining to that problem.

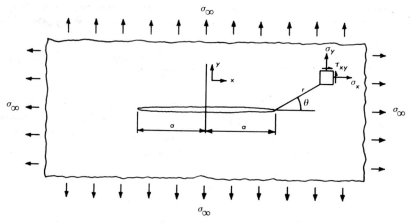

Figure 2.2. A biaxially loaded infinite plate containing a slit crack.

For an infinite, biaxially loaded plate containing a crack as shown in figure 2.2 the boundary conditions are readily stated, namely

1) $\sigma_y = 0$ for $-a < x < +a$ and $y = 0$
2) $\sigma_y \to \sigma_\infty$ as $x \to \pm\infty$
3) $\sigma_y \to \infty$ (i.e. a singularity) at $x = \pm a$, since the crack is a stress raiser.

Now a function $\phi(z)$ that fulfils these boundary conditions must be chosen. One example is the function

$$\phi(z) = \frac{\sigma}{\sqrt{1 - a^2/z^2}} . \tag{2.6}$$

We know from equation (2.5) that along the x-axis (i.e. $y = 0$) $\sigma_y = \mathrm{Re}\,\phi(z)$, and it is seen from equation (2.6) that for $-a < x < +a$ $\mathrm{Re}\,\phi(z) = 0$ and hence $\sigma_y = 0$, as required by the first boundary condition. Furthermore, for $z \to \infty$ equation (2.6) gives $\phi(z) = \sigma$, and for $z = a$ or $-a$ there is a singularity, so that the second and third boundary conditions are also fulfilled.

However, if the origin is translated to the crack tip by taking $\eta = z - a$, we obtain another suitable stress function which turns out to be easier to use. This function is

$$\phi(\eta) = \frac{\sigma}{\sqrt{1 - a^2/(a+\eta)^2}} = \frac{\sigma(a+\eta)}{\sqrt{(a+\eta)^2 - a^2}} . \tag{2.7}$$

A first order approximation for this equation, valid for $\eta \ll a$, is simply

$$\phi(\eta) = \frac{\sigma a}{\sqrt{2a\eta}} = \sigma \sqrt{\frac{a}{2\eta}} . \tag{2.8}$$

Note that this approximation is the first term in a series expansion. Using equation (2.8) a change can easily be made to a polar coordinate representation (r, θ) with origin at the crack tip as shown in figure 2.2. Then $\eta = re^{i\theta}$ and

$$\phi(\eta) = \frac{\sigma\sqrt{a}}{\sqrt{2r}} \cdot e^{-\frac{1}{2}i\theta} \quad \text{or} \quad \frac{\sigma\sqrt{\pi a}}{\sqrt{2\pi r}} \cdot e^{-\frac{1}{2}i\theta} . \tag{2.9}$$

Equation (2.9) is only valid for $r \ll a$, owing to the approximation made to obtain equation (2.8). Having obtained equation (2.9) straightforward algebra can be used to find $\mathrm{Re}\,\phi(\eta)$, $\mathrm{Re}\,\phi'(\eta)$ and $\mathrm{Im}\,\phi'(\eta)$. In turn these quantities can be substituted into equations (2.5), resulting in

$$\sigma_x = \frac{\sigma\sqrt{\pi a}}{\sqrt{2\pi r}} \cos\frac{\theta}{2} \left(1 - \sin\frac{\theta}{2}\sin\frac{3\theta}{2}\right)$$

$$\sigma_y = \frac{\sigma\sqrt{\pi a}}{\sqrt{2\pi r}} \cos\frac{\theta}{2} \left(1 + \sin\frac{\theta}{2}\sin\frac{3\theta}{2}\right) \tag{2.10}$$

$$\tau_{xy} = \frac{\sigma\sqrt{\pi a}}{\sqrt{2\pi r}} \sin\frac{\theta}{2}\cos\frac{\theta}{2}\cos\frac{3\theta}{2} .$$

Equations (2.10) show that all the stresses tend to infinity at $r = 0$ (the crack tip) and are products of the geometrical position $\frac{1}{\sqrt{2\pi r}}$ $f(\theta)$ and a factor $\sigma\sqrt{\pi a}$, which is a simple function of remote stress and crack length. Thus the factor $\sigma\sqrt{\pi a}$ determines the magnitude of the elastic stresses in the crack tip field.

This factor is called the mode I *stress intensity factor*, $K_I = \sigma\sqrt{\pi a}$.

Remarks

1) The foregoing derivation is only one of the methods for obtaining the stress field solution. There are several other, more general methods.
2) The derivation is specific to a biaxially loaded infinite plate. However, it can be shown that for a uniaxially loaded plate the solution differs from equation (2.10) only in that a non-singular term σ_∞ must be subtracted from the expression for σ_x. This correction is often omitted, since the first term in the expression is much larger.
3) The solution is only the first term in a series expansion and is valid only for $\eta, r \ll a$. However, this restriction leads to a very important result. For every mode I problem the general form of $\phi(z)$ must contain the factor $\dfrac{1}{\sqrt{1 - a^2/z^2}}$ to ensure that $\sigma_y = 0$ for $-a < x < +a$. Then performing the $\eta = z - a$ transition means that for $\eta \ll a$ all stress functions $\phi(\eta)$ reduce to

$$\phi(\eta) = \frac{f(\eta)}{\sqrt{\eta}} \, .$$

Moreover, because $\eta \ll a$, it may be considered to tend to zero and $f(\eta)$ can be replaced by a constant value. Choosing this value to be $K_I/\sqrt{2\pi}$, we may write

$$\lim_{\eta \to 0} \phi(\eta) = \frac{K_I}{\sqrt{2\pi\eta}} \, .$$

Comparing this result with equation (2.9) it is seen that for every mode I stress function the geometric part $f(r,\theta)$ remains the same and only K_I varies. The consequence is that in the vicinity of the crack tip ($\eta \ll a$) the total stress field due to two or more different mode I loading systems can be obtained by simple algebraic summation of the respective stress intensity factors. This is a very important result, as will be shown in section 2.6.

4) The solution is valid only for a slit-shaped crack with zero crack tip radius. In practice it is sometimes necessary to consider a crack with finite tip radius. A solution to this problem was obtained by Creager and Paris, who simply moved the origin of the polar coordinate system an amount equal to one half the crack tip radius, as shown in figure 2.3.

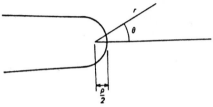

Figure 2.3. Crack with a finite tip radius, ρ.

The following expressions were obtained:

$$\sigma_x = \frac{K_I}{\sqrt{2\pi r}} \cos\frac{\theta}{2} \left(1 - \sin\frac{\theta}{2}\sin\frac{3\theta}{2}\right) - \frac{K_I}{\sqrt{2\pi r}}\left(\frac{\rho}{2r}\right)\cos\frac{3\theta}{2}$$

$$\sigma_y = \frac{K_I}{\sqrt{2\pi r}} \cos\frac{\theta}{2} \left(1 + \sin\frac{\theta}{2}\sin\frac{3\theta}{2}\right) + \frac{K_I}{\sqrt{2\pi r}}\left(\frac{\rho}{2r}\right)\cos\frac{3\theta}{2} \quad (2.11)$$

$$\tau_{xy} = \frac{K_I}{\sqrt{2\pi r}} \sin\frac{\theta}{2} \cos\frac{\theta}{2} \cos\frac{3\theta}{2} - \frac{K_I}{\sqrt{2\pi r}}\left(\frac{\rho}{2r}\right)\sin\frac{3\theta}{2}.$$

Note that since r is finite at the crack tip ($r = \rho/2$) the stresses are finite and there is no crack tip singularity as in the case of a slit crack.

5) The derivation so far has only considered a state of plane stress, i.e. $\sigma_z = 0$. For plane strain the elastic stress $\sigma_z = \nu(\sigma_x + \sigma_y)$, where ν is Poisson's ratio. Thus σ_z can be calculated from the expressions for σ_x and σ_y in equations (2.10).

2.3. Useful Expressions

In this section some useful expressions for stresses and strains in the crack tip region will be given, namely

- The mode I stress field equations in terms of principal stresses
- The stress field equations for modes II and III
- The elastic stress field displacements.

Expressing stress field equations in terms of principal stresses is useful when considering yield criteria in order to estimate plastic zone sizes, as will be discussed in chapter 3. Stress field equations for modes II and III are required for studying crack problems in which these modes of loading or combined mode loading apply. Expressions for elastic stress field displacements enable calculation of the stored elastic energy (used in energy balance approaches, chapter 4), and also serve as a basis for displacement controlled fracture criteria, e.g. COD (chapter 7).

Mode I Stress Field Equations in Terms of Principal Stresses

The derivation is obtained from a Mohr's circle construction, figure 2.4.

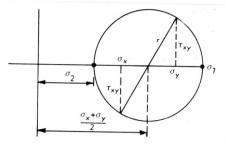

Figure 2.4. Mohr's circle construction.

From this figure it is seen that

$$\sigma_{1,2} = \frac{\sigma_x + \sigma_y}{2} \pm r$$

where

$$r = \sqrt{\left(\frac{\sigma_y - \sigma_x}{2}\right)^2 + \tau_{xy}^2}$$

so that

$$\sigma_{1,2} = \frac{\sigma_x + \sigma_y}{2} \pm \sqrt{\left(\frac{\sigma_y - \sigma_x}{2}\right)^2 + \tau_{xy}^2} . \qquad (2.12)$$

Substitution of equations (2.10) for σ_x, σ_y and τ_{xy} in equation (2.12) gives

$$\sigma_1 = \frac{K_I}{\sqrt{2\pi r}} \cos\frac{\theta}{2} \left(1 + \sin\frac{\theta}{2}\right)$$

$$\sigma_2 = \frac{K_I}{\sqrt{2\pi r}} \cos\frac{\theta}{2} \left(1 - \sin\frac{\theta}{2}\right). \qquad (2.13)$$

Since σ_3 is either 0 (plane stress) or $\nu(\sigma_1 + \sigma_2)$ (plane strain), then for plane strain

$$\sigma_3 = \frac{2\nu K_I}{\sqrt{2\pi r}} \cos\frac{\theta}{2} .$$

The Stress Field Equations for Modes II and III

The stress field equations for mode II and mode III loading may be obtained in a similar manner to that outlined in section 2.2 for mode I. The results are:

$$\text{Mode II} \atop K_{II} = \tau \sqrt{\pi a} \quad \begin{cases} \sigma_x = \dfrac{-K_{II}}{\sqrt{2\pi r}} \sin\dfrac{\theta}{2} (2 + \cos\dfrac{\theta}{2} \cos\dfrac{3\theta}{2}) \\[3mm] \sigma_y = \dfrac{K_{II}}{\sqrt{2\pi r}} \sin\dfrac{\theta}{2} (\cos\dfrac{\theta}{2} \cos\dfrac{3\theta}{2}) \\[3mm] \tau_{xy} = \dfrac{K_{II}}{\sqrt{2\pi r}} \cos\dfrac{\theta}{2} (1 - \sin\dfrac{\theta}{2} \sin\dfrac{3\theta}{2}). \end{cases} \qquad (2.14)$$

$$\text{Mode III} \atop K_{III} = \tau \sqrt{\pi a} \quad \begin{cases} \tau_{xz} = \dfrac{-K_{III}}{\sqrt{2\pi r}} \sin\dfrac{\theta}{2} \\[3mm] \tau_{yz} = \dfrac{-K_{III}}{\sqrt{2\pi r}} \cos\dfrac{\theta}{2}. \end{cases} \qquad (2.15)$$

The Elastic Stress Field Displacements

In a completely elastic situation Hooke's law applies, i.e. for plane strain

$$E\epsilon_x = \sigma_x - \nu(\sigma_y + \sigma_z)$$
$$= (1 - \nu^2)\sigma_x - \nu(1 + \nu)\sigma_y. \qquad (2.16)$$

Now $\epsilon_x = \dfrac{\partial u}{\partial x}$, where u is the elastic displacement in the x direction, so that

$$u = \frac{1}{E} \int [(1 - \nu^2)\sigma_x - \nu(1 + \nu)\sigma_y] \, dx. \qquad (2.17)$$

The case of plane strain is chosen because it can be proved that the integration constant for equation (2.17) is zero. Substituting for σ_x and σ_y from equations (2.5) and integrating equation (2.17) results in

$$u = (\frac{1 + \nu}{E}) [(1 - 2\nu) \operatorname{Re} \overline{\phi}(z) - y \operatorname{Im} \phi(z)]. \qquad (2.18.a)$$

Similarly, for $\epsilon_y = \dfrac{\partial v}{\partial y}$ one can obtain v, the elastic displacement in y direction, by integrating with the help of the Cauchy-Rieman equations, equations (2.4). Then

$$v = (\frac{1 + \nu}{E}) [2(1 - \nu) \operatorname{Im} \overline{\phi}(z) - y \operatorname{Re} \phi(z)]. \qquad (2.18.b)$$

$\operatorname{Re} \overline{\phi}(z)$, $\operatorname{Im} \overline{\phi}(z)$, $\operatorname{Im} \phi(z)$ and $\operatorname{Re} \phi(z)$ can be obtained from equation (2.9). Substitution into equations (2.18) gives

$$\text{plane} \atop \text{strain} \quad \begin{cases} u = 2(1+\nu)\dfrac{K_I}{E}\sqrt{\dfrac{r}{2\pi}}\ \cos\dfrac{\theta}{2}(2-2\nu-\cos^2\dfrac{\theta}{2}) \\[4mm] v = 2(1+\nu)\dfrac{K_I}{E}\sqrt{\dfrac{r}{2\pi}}\ \sin\dfrac{\theta}{2}(2-2\nu-\cos^2\dfrac{\theta}{2}). \end{cases} \qquad (2.19)$$

Note that the only differences between u and v are the factors $\cos\theta/2$ and $\sin\theta/2$.

For plane stress the derivation of elastic stress field displacements is more complex. Only the result is given here:

$$\text{plane} \atop \text{stress} \quad \begin{cases} u = 2\dfrac{K_I}{E}\sqrt{\dfrac{r}{2\pi}}\ \cos\dfrac{\theta}{2}(1+\sin^2\dfrac{\theta}{2}-\nu\cos^2\dfrac{\theta}{2}) \\[4mm] v = 2\dfrac{K_I}{E}\sqrt{\dfrac{r}{2\pi}}\ \sin\dfrac{\theta}{2}(1+\sin^2\dfrac{\theta}{2}-\nu\cos^2\dfrac{\theta}{2}). \end{cases} \qquad (2.20)$$

Another useful expression is the formula for displacement v of a crack flank at any position away from the crack tip. Figure 2.5 shows a schematic definition of crack flank displacement.

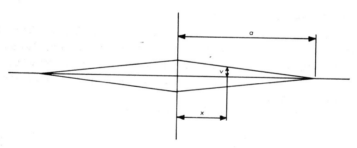

Figure 2.5. Crack flank displacement.

Since $\mathrm{Re}\,\phi(z) = 0$ for $-a < x < +a$ and $y = 0$, from equation (2.18.b) it follows that

$$\text{plane} \atop \text{strain} \quad v = \frac{2(1-\nu^2)}{E}\ \mathrm{Im}\,\overline{\phi}(z) = \frac{2\sigma}{E}(1-\nu^2)\sqrt{a^2-x^2}. \qquad (2.21)$$

For plane stress the factor $(1-\nu^2)$ disappears, i.e.

$$\text{plane} \atop \text{stress} \quad v = \frac{2\sigma}{E}\sqrt{a^2-x^2}. \qquad (2.22)$$

The disappearance of $(1-\nu^2)$ follows from the change in the expression for Hooke's law in going from plane strain to plane stress. If all equations are expressed in terms of the sliding modulus, $\mu = E/2(1+\nu)$, thereby taking the sliding deformation γ_{xy} into account, Hooke's law in terms of μ is:

a) Plane Strain

$$\epsilon_x = \frac{1-\nu}{2\mu} \sigma_x - \frac{\nu}{2\mu} \sigma_y$$

$$\epsilon_y = \frac{1-\nu}{2\mu} \sigma_y - \frac{\nu}{2\mu} \sigma_x .$$

b) Plane Stress

$$\epsilon_x = \frac{1}{2\mu(1+\nu)} (\sigma_x - \nu\sigma_y)$$

$$\epsilon_y = \frac{1}{2\mu(1+\nu)} (\sigma_y - \nu\sigma_x).$$

Note that when ν is replaced by $\nu/(1+\nu)$ in the equations for plane strain then the plane stress equations are obtained.

The plane strain displacement in terms of μ is

plane strain
$$v = \frac{\sigma(1-\nu)}{\mu} \sqrt{a^2 - x^2}. \qquad (2.21.a)$$

Replacing ν by $\nu/(1+\nu)$ gives the plane stress displacement

plane stress
$$v = \frac{2\sigma}{2\mu(1+\nu)} \sqrt{a^2 - x^2} \qquad (2.22.a)$$

and since $\mu = E/2(1+\nu)$ the plane stress displacement in terms of E is

plane stress
$$v = \frac{2\sigma}{E} \sqrt{a^2 - x^2}.$$

2.4. Finite Specimen Width

The solution for the stress intensity factor in section 2.2 is strictly valid only for an infinite plate. The geometry of finite size specimens has an effect on the crack tip stress field, and so expressions for stress intensity factors have to be modified by the addition of correction factors to enable their use in practical problems.

A general form for such a modified expression is

$$K_I = C\sigma\sqrt{\pi a} \cdot f(\frac{a}{W}) \qquad (2.23)$$

where C and $f(\frac{a}{W})$ have to be determined by stress analysis. There are very few closed form solutions to equation (2.23). Most expressions are obtained by numerical approximation methods.

As illustrations of the effect of finite geometry the derivation of modified expressions based on the solution for an infinite, remotely loaded plate will be discussed in some detail. These expressions concern the centre cracked specimen and the single and double edge notched specimens. (A compendium of these and additional expressions for a number of well known specimen geometries is given in section 2.8.)

The Centre Cracked Specimen

The specimen geometry is depicted in figure 2.6. For this specimen there are several expressions for the stress intensity factor, e.g.

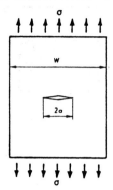

Figure 2.6. The finite width centre cracked specimen.

$$K_I = \sigma \sqrt{\pi a} \ \sqrt{\frac{W}{\pi a}} \ \tan(\frac{\pi a}{W}).$$ (2.24)

Equation (2.24) is the analytically obtained stress intensity factor for one of a row of collinear cracks with interspacing W in an infinite plate. The accuracy is better than 5% for $a/W \leqslant 0.25$.

A virtually exact numerical solution was obtained by Isida. The geometric correction factor $f(\frac{a}{W})$ was derived as a 36 term power series! However, Brown found a 4 term approximation to this power series with 0.5% accuracy for $a/W \leqslant 0.35$. This approximation is

$$f(\frac{a}{W}) = 1 + 0.256\,(\frac{a}{W}) - 1.152\,(\frac{a}{W})^2 + 12.200\,(\frac{a}{W})^3:$$ (2.25)

Another, purely empirical, correction factor is due to Feddersen. As an approximation to Isida's results he suggested that

$$K_I = \sigma \sqrt{\pi a} \ \sqrt{\sec(\frac{\pi a}{W})}.$$ (2.26)

This remarkably simple expression is accurate to within 0.3% for $a/W \leqslant 0.35$.

A more recent correction factor introduced by Dixon is often used:

$$K_I = \sigma \sqrt{\pi a} \ \frac{1}{\sqrt{1 - (\frac{2a}{W})^2}}.$$ (2.27)

Figure 2.7 shows all these correction factors in a graphical representation for comparison.

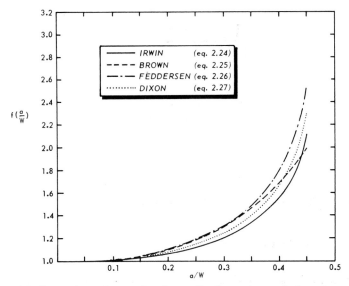

Figure 2.7. Comparison of correction factors for the centre cracked specimen.

Edge Notched Specimens

The geometries of single and double edge notched specimens are given in figure 2.8.

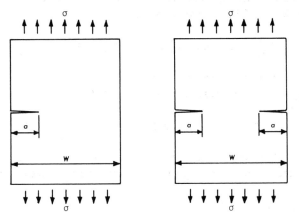

Figure 2.8. The single and double edge notched specimens.

In remark (2) appended to section 2.2 it was stated that a non-singular term σ_∞ must be subtracted from equation (2.10) in order to obtain σ_x for a uniaxially loaded plate. Usually σ_∞ can be neglected. However, for $-a < x < +a$ equation (2.10) gives a stress of zero in the negative x direction ($\theta = 180°$) and σ_∞ is no longer negligible. There are thus compressive stresses

$\sigma_x = -\sigma_\infty$ acting along the crack surfaces.

These compressive stresses have a closing effect on the crack in a centre cracked specimen. This closing effect is absent in the case of edge cracks, since σ_x at the free edge is zero. Thus there is a stress raising effect of the free edge. This effect is at a maximum for short edge cracks $(f(\frac{a}{W}) \sim 1)$ and has been estimated to be about 12%, i.e.

$$K_I = 1.12\,\sigma\sqrt{\pi a}. \qquad (2.28)$$

For longer cracks the finite geometry results in stress enhancement whereas the free edge effect diminishes. Correction factors have to take these trends into account, and this is reflected in the following most accurate expressions.

single edge notched specimen

$$K_I = \sigma\sqrt{\pi a}\;[1.12 - 0.231\,(\tfrac{a}{W}) + 10.55\,(\tfrac{a}{W})^2 - 21.72\,(\tfrac{a}{W})^3 + 30.39\,(\tfrac{a}{W})^4]$$

$$(2.29)$$

which has an accuracy of 0.5% for a/W < 0.6.

double edge notched specimen

$$K_I = \sigma\sqrt{\pi a}\;[\frac{1.122 - 0.561\,(\tfrac{a}{W}) - 0.205\,(\tfrac{a}{W})^2 + 0.471\,(\tfrac{a}{W})^3 - 0.190\,(\tfrac{a}{W})^4}{\sqrt{1 - \tfrac{a}{W}}}]$$

$$(2.30)$$

which is accurate to 0.5% for any a/W < 0.5.

2.5. Two Additional Important Solutions for Practical Use

Besides the centre cracked and single and double edge notched geometries there are two other geometries which are important owing to their common occurrence in practice. These are:
- Crack line loading
- Elliptical cracks, either embedded or intersecting free surfaces.

Elliptical cracks at free surfaces may be semi or quarter elliptical in shape.

Crack Line Loading

Consider a crack loaded by a point force as in figure 2.9. This is known as crack line loading.

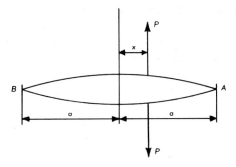

Figure 2.9. Crack line loading by a point force P.

Expressions for the stress intensity factors at crack tips A and B can be obtained by a similar analysis to that in section 2.2. The result in terms of P per unit thickness is:

$$K_{I_A} = \frac{P}{\sqrt{\pi a}} \sqrt{\frac{a + x}{a - x}} \quad ; \quad K_{I_B} = \frac{P}{\sqrt{\pi a}} \sqrt{\frac{a - x}{a + x}} . \qquad (2.31.a)$$

For a centrally located force (x = 0)

$$K_I = \frac{P}{\sqrt{\pi a}} . \quad \text{(per unit thickness)} \qquad (2.31.b)$$

Therefore for constant P an increase in crack length results in a decrease in stress intensity.

From equations (2.31.a) a stress intensity factor solution for a crack under internal pressure (= P per unit area) can be derived, as follows.

$$K_I = \frac{P}{\sqrt{\pi a}} \int_0^a \left(\sqrt{\frac{a + x}{a - x}} + \sqrt{\frac{a - x}{a + x}} \right) dx$$

$$= \frac{P}{\sqrt{\pi a}} \int_0^a \left(\sqrt{\frac{(a + x)^2}{a^2 - x^2}} + \sqrt{\frac{(a - x)^2}{a^2 - x^2}} \right) dx$$

$$= \frac{P}{\sqrt{\pi a}} \int_0^a \left(\frac{a + x}{\sqrt{a^2 - x^2}} + \frac{a - x}{\sqrt{a^2 - x^2}} \right) dx$$

$$= \frac{P}{\sqrt{\pi a}} \int_0^a \frac{2a \, dx}{\sqrt{a^2 - x^2}} .$$

Substituting x = a cos φ,

$$K_I = 2P \sqrt{\frac{a}{\pi}} \int_0^a \frac{da \cos \varphi}{\sqrt{a^2 - a^2 \cos^2 \varphi}} = 2P \sqrt{\frac{a}{\pi}} \int_0^a \frac{da \cos \varphi}{a \sin \varphi}$$

$$= 2P \sqrt{\frac{a}{\pi}} \; [-\varphi]_0^a = 2P \sqrt{\frac{a}{\pi}} \; [-\arccos\left(\frac{x}{a}\right)]_0^a$$

$$= P\sqrt{\pi a}. \tag{2.32}$$

Note that since P is for unit area (pressure) the result in equation (2.32) is the same as that obtained by end loading, $K_I = \sigma\sqrt{\pi a}$.

The usefulness of solutions for crack line loading is twofold:

1) Expressions for point forces can be applied to cracks at loaded holes, e.g. riveted and bolted plates, provided that the holes are not too large with respect to the crack.

2) Integration of point force solutions to obtain expressions for cracks under internal pressure is particularly useful for analysing internal part-through wall thickness cracks in e.g. pressure vessels and piping.

Elliptical Cracks

Actual cracks often initiate at surface discontinuities or corners in structural components. If the components are fairly thick the cracks generally assume semi or quarter elliptical shapes as they grow in the thickness direction, for example as in figure 2.10. A knowledge of stress intensity factors for these geometries is thus of prime importance for practical application of LEFM.

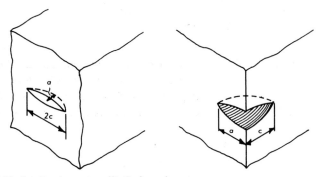

Figure 2.10. Semi and quarter elliptical cracks.

As a first step, Irwin derived an expression for the mode I stress intensity factor of an embedded slit-like elliptical crack. The loading situation is shown schematically in figure 2.11.

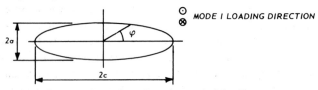

Figure 2.11. An embedded elliptical crack under mode I loading.

For this situation

$$K_I = \frac{\sigma\sqrt{\pi a}}{\Phi} (\sin^2\varphi + \frac{a^2}{c^2} \cos^2\varphi)^{\frac{1}{4}} \qquad (2.33)$$

where Φ is an elliptic integral of the second kind, i.e. $\int_0^{\frac{\pi}{2}} \sqrt{1 - \sin^2\alpha \sin^2\varphi} \, d\varphi$

and $\sin^2\alpha = (c^2 - a^2)/c^2$. The solution for the integral is given in the table below for a number of a/c ratios.

a/c	0	0.1	0.2	0.3	0.4	0.5	0.6	0.7	0.8	0.9	1.0
Φ	1.000	1.016	1.051	1.097	1.151	1.211	1.277	1.345	1.418	1.493	1.571

The integral can also be developed as a series expansion:

$$\Phi = \frac{\pi}{2}[1 - \frac{1}{4}\frac{c^2 - a^2}{c^2} - \frac{3}{64}(\frac{c^2 - a^2}{c^2})^2 - \ldots].$$

Neglecting all terms beyond the second gives an accuracy of better than 5%. And so the stress intensity factor can be approximated by

$$K_I = \frac{\sigma\sqrt{\pi a}}{\frac{3\pi}{8} + \frac{\pi}{8}\frac{a^2}{c^2}} (\sin^2\varphi + \frac{a^2}{c^2} \cos^2\varphi)^{\frac{1}{4}}. \qquad (2.34)$$

K_I varies along the elliptical crack front and has a maximum value $\sigma\sqrt{\pi a}/\Phi$ at the ends of the minor axis, a, and a minimum value $\sigma\sqrt{\pi a^2/c}/\Phi$ at the ends of the major axis, c. The implication is that during crack growth an embedded elliptical crack will tend to become circular. For a circular crack

$$K_I = \frac{2}{\pi} \sigma\sqrt{\pi a}. \qquad (2.35)$$

In practice elliptical cracks will generally occur as semi elliptical surface flaws or quarter elliptical corner cracks. The presence of free surfaces means that correction factors must be added to the expressions given for embedded cracks.

For a semi elliptical surface flaw the free front surface is generally accounted for by adding a correction factor of 1.12. For a quarter elliptical corner crack, which intersects two surfaces, a correction factor of 1.2 is used. However, if surface cracks grow deeper into the material the surface correction factors decrease from 1.12 or 1.2 to 1. Besides these factors, corrections for cracks approaching back surfaces (analogous to corrections for finite specimen width, section 2.4) must be considered. Back surface correction factors have been calculated by various authors: an overview can be found in reference 3 of the bibliography. The results are often combined with front free surface and finite width correction factors.

The best solutions available for semi elliptical surface cracks are those based on the finite element calculations of Raju and Newman, reference 4 of the bibliography.

For the crack of figure 2.12 Raju and Newman produced stress intensity factor solutions of the form $K = C\sigma\sqrt{\pi a}/\Phi$ where, for $W \gg c$ the value of C depends only on a/c, a/B and φ. Values of C are also given in figure 2.12.

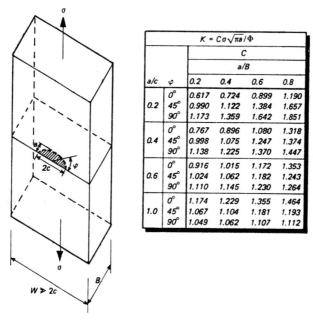

$K = C\sigma\sqrt{\pi a}/\Phi$					
		C			
		a/B			
a/c	φ	0.2	0.4	0.6	0.8
0.2	0°	0.617	0.724	0.899	1.190
	45°	0.990	1.122	1.384	1.657
	90°	1.173	1.359	1.642	1.851
0.4	0°	0.767	0.896	1.080	1.318
	45°	0.998	1.075	1.247	1.374
	90°	1.138	1.225	1.370	1.447
0.6	0°	0.916	1.015	1.172	1.353
	45°	1.024	1.062	1.182	1.243
	90°	1.110	1.145	1.230	1.264
1.0	0°	1.174	1.229	1.355	1.464
	45°	1.067	1.104	1.181	1.193
	90°	1.049	1.062	1.107	1.112

Figure 2.12. Stress intensity factor solutions for semi elliptical surface cracks in a plate of finite dimensions according to Raju and Newman (reference 4).

Raju and Newman also developed an empirical stress intensity factor equation based on their finite element results, reference 5 of the bibliography. For tension loading the solution is reproduced in section 2.8. For combined tension and bending load the reader is referred to reference 5.

2.6. Superposition of Stress Intensity Factors

In remark (3) appended to section 2.2 it was shown that in the vicinity of the crack tip the total stress field due to two or more different mode I loading systems can be obtained by an algebraic summation of the respective stress intensity factors. This is called the superposition principle.

It should be noted that the superposition principle is valid only for combinations of the same mode of loading, i.e. all mode I, all mode II or all mode III. The different modes give different types of stress intensity factor solutions which cannot be superimposed.

By using the superposition principle the stress intensity factor for a number of seemingly complicated problems can be readily obtained. Two examples are given here. They are:

1) A through crack under internal pressure (already derived analytically in section 2.5).

2) A semi elliptical surface flaw in a cylindrical pressure vessel.

Through Crack under Internal Pressure

The solution is obtained by the superposition shown in figure 2.13. It is seen from this superposition that

$$K_I^A = K_I^B = K_I^C + K_I^D = 0.$$

Therefore

$$K_I^D = -K_I^C = -\sigma\sqrt{\pi a}.$$

Reversal of the stress direction in D gives the required result, $\sigma\sqrt{\pi a}$.

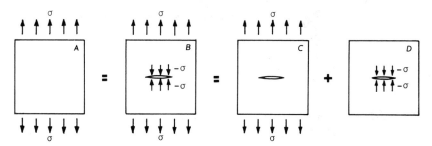

Figure 2.13. The superposition principle for a through crack under internal pressure.

Semi Elliptical Surface Flaw in a Cylindrical Pressure Vessel

A section from a cylindrical pressure vessel with an internal semi elliptical surface flaw is given in figure 2.14. As long as the wall thickness, B, is small compared to the vessel radius, R, the hoop stress σ_H = PR/B. The maximum stress intensity factor will generally occur at the end of the minor axis of the semi elliptical surface flaw ($\varphi = 90°$). The contribution to this maximum stress intensity factor by the hoop stress is

$$K_I^{\sigma_H} = \frac{C\sigma_H\sqrt{\pi a}}{\Phi} = \frac{CPR\sqrt{\pi a}}{B\Phi} \tag{2.36}$$

where C can be obtained from figure 2.12 when a and c are known.

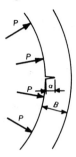

Figure 2.14. Cylindrical pressure vessel with internal surface flaw.

Since the crack is in a pressurized vessel the internal pressure will also act on the crack surfaces, with a contribution to the maximum stress intensity factor of

$$K_I^P = \frac{CP\sqrt{\pi a}}{\Phi}. \qquad (2.37)$$

The maximum stress intensity factor is therefore

$$K_{Imax} = K_I^{\sigma H} + K_I^P = \frac{CP(1 + \frac{R}{B})\sqrt{\pi a}}{\Phi}. \qquad (2.38)$$

2.7. Some Remarks Concerning Stress Intensity Factor Determinations

In attempting to use fracture mechanics it will often be found that there is no standard stress intensity factor solution for the particular crack shape and structural component geometry under consideration. Recourse must be made to one of a number of methods for obtaining the stress intensity factor. A detailed treatise of these methods is beyond the scope of this course, but some remarks on the subject will be made here.

Stress intensity factors are now available for many geometrical configurations, and so the first step should always be a literature search. For example, references 6 and 7 of the bibliography to this chapter and the indexes of well known journals such as the "International Journal of Fracture" and "Engineering Fracture Mechanics" can serve as starting points.

If no applicable solution is directly available, the next step is to assess the permissible effort to solve the problem. This effort depends on the seriousness of the problem, the desired accuracy, computational costs and how many times the solution will be useful. Limits to the amount of effort that can be justified will frequently rule out the use of sophisticated and expensive methods such as

- Finite element calculations

- Boundary integral equations
- Conformal mapping.

The interested reader will find detailed information about these methods in reference 8 of the bibliography. A strong mathematical background is essential for using these methods.

However, in many cases one of several more straightforward methods can be applied. These include

- The superposition principle, already discussed in section 2.6; or the very similar compounding method, reference 9 of the bibliography
- The experimental fatigue crack growth method
- The weight function method.

The Experimental Fatigue Crack Growth Method

This method generally gives reliable results. It involves comparing fatigue crack growth rates in the configuration to be investigated with crack growth rates in standard specimens of the same material fatigued under exactly the same conditions. Constant amplitude loading is used, since crack growth rate data are then correlatable by ΔK, the stress intensity factor range, as mentioned in chapter 1.

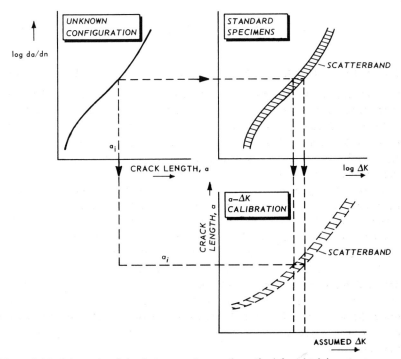

Figure 2.15. Schematic of the fatigue crack growth method for obtaining stress intensity factors.

In order to determine the stress intensity factor the assumption is made that at the same crack growth rate da/dn, the same ΔK applies to the standard specimens and the configuration being considered. Fatigue crack propagation tests over a range of crack lengths result in ΔK (and hence K_I) calibration curves, as shown schematically in figure 2.15.

The Weight Function Method

From expressions for the elastic energy release rate, G, and the relation $G = K^2/E$ (see chapter 4) it can be shown that the stress intensity factor for a particular stress distribution can be calculated if the stress intensity factor is known for another stress distribution on the same configuration of specimen and crack geometries. The proof of this is given in reference 10 of the bibliography. The result for mode I loading is

$$K_I = \int_0^{2a} H(x,a) \cdot \sigma(x)\,dx \tag{2.39}$$

where the weight function

$$H(x,a) = \frac{E'}{2K_I^*} \frac{\partial v^*(x,a)}{\partial a}.$$

E' is E for plane stress, and $E/(1-\nu^2)$ for plane strain; K_I^* is the known stress intensity factor; v^* is the displacement at the loading point, x, of the known solution but in the direction of loading for the stress distribution to be analysed; and $\sigma(x)$ is a function describing this stress distribution with reference to the stress distribution for K_I^*.

A serious difficulty in using equation (2.39) is the part $\int \frac{\partial v^*}{\partial a}\,dx$, which often gives singularities in the stress field distribution around a crack tip. A better technique is to use special weight function solutions which give singularities only at the crack tip. Such special weight functions have been derived by Bueckner and, in a different way, by Paris. Discussion of these functions, including examples of their use, can be found in references 2 and 10 of the bibliography to this chapter.

2.8. A Compendium of Well Known Stress Intensity Factor Solutions

A number of well known and widely used stress intensity factor solutions are presented here. Some of these solutions have already been discussed in sections 2.4 – 2.6. Others will be found in subsequent chapters of this course or else are mentioned because of their practical utility.

Elementary Solutions

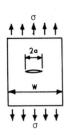

Centre cracked plate: $K_I = C\sigma\sqrt{\pi a}$

$$C = 1 + 0.256\left(\frac{a}{W}\right) - 1.152\left(\frac{a}{W}\right)^2 + 12.200\left(\frac{a}{W}\right)^3$$

or $C = \sqrt{\sec\left(\frac{\pi a}{W}\right)}$ (Feddersen)

or $C = \dfrac{1}{\sqrt{1 - \left(\frac{2a}{W}\right)^2}}$ (Dixon)

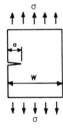

Single edge notched plate: $K_I = C\sigma\sqrt{\pi a}$

$C = 1.12$ for small cracks

or $C = 1.12 - 0.231\left(\frac{a}{W}\right) + 10.55\left(\frac{a}{W}\right)^2 - 21.72\left(\frac{a}{W}\right)^3 + 30.39\left(\frac{a}{W}\right)^4$

up to $a/W = 0.6$

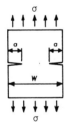

Double edge notched plate: $K_I = C\sigma\sqrt{\pi a}$

$C = 1.12$ for small cracks

or $C = \dfrac{1.122 - 0.561\left(\frac{a}{W}\right) - 0.205\left(\frac{a}{W}\right)^2 + 0.471\left(\frac{a}{W}\right)^3 - 0.190\left(\frac{a}{W}\right)^4}{\sqrt{1 - \frac{a}{W}}}$

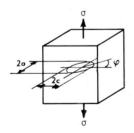

Embedded elliptical or circular slit crack:

Elliptical crack $K_I = \dfrac{\sigma\sqrt{\pi a}}{\frac{3\pi}{8} + \frac{\pi}{8}\frac{a^2}{c^2}}\left(\sin^2\varphi + \frac{a^2}{c^2}\cos^2\varphi\right)^{\frac{1}{4}}$

Circular crack $K_I = \dfrac{2}{\pi}\sigma\sqrt{\pi a}$

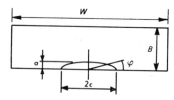

Semi elliptical surface flaw in tension:

$$K_I = C \frac{\sigma \sqrt{\pi a}}{\frac{3\pi}{8} + \frac{\pi}{8} \frac{a^2}{c^2}} (\sin^2 \varphi + \frac{a^2}{c^2} \cos^2 \varphi)^{\frac{1}{4}}$$

C = 1.12   for shallow cracks

or C = correction factor for K_I to be obtained from figure 2.12

or: $K_I = C\sigma \dfrac{\sqrt{\pi a}}{\Phi}$

$$C = f(\frac{a}{B}, \frac{a}{c}, \frac{c}{W}, \varphi)$$

$$C = [C_1 + C_2(\frac{a}{B})^2 \pm C_3(\frac{a}{B})^4] C_4 g(\varphi) g(w)$$

$$C_1 = 1.13 - 0.09(\frac{a}{c})$$

$$C_2 = -0,54 + \frac{0.89}{0.2 + (\frac{a}{c})}$$

$$C_3 = 0.5 - \frac{1.0}{0.65 + (\frac{a}{c})} + 14(1.0 - \frac{a}{c})^{24}$$

$$C_4 = 1 + [0.1 + 0.35(\frac{a}{B})^2](1 - \sin\varphi)^2$$

$$g(\varphi) = [\sin^2\varphi + (\frac{a}{c})^2 \cos^2\varphi]^{\frac{1}{4}}$$

$$g(W) = [\sec\frac{\pi c}{w} \sqrt{\frac{a}{B}}]^{\frac{1}{2}}$$

for: $0 < \dfrac{a}{c} < 1$ $0 < \dfrac{a}{B} \leqslant 1$

$0 < \dfrac{c}{B} < 0.5$ $0 < \varphi < \pi$

Quarter elliptical corner crack in tension:

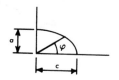

$$K_I = C \frac{\sigma \sqrt{\pi a}}{\frac{3\pi}{8} + \frac{\pi}{8} \frac{a^2}{c^2}} (\sin^2 \varphi + \frac{a^2}{c^2} \cos^2 \varphi)^{\frac{1}{4}}$$

C = 1.2

The solutions for corner cracks in finite specimens and for corner cracks at holes are not elementary. They are, however, very important and the reader is referred to references 11 and 12 of the bibliography.

Solutions for Standard Test Specimens

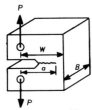

Compact tension specimen (CT):

$$K_I = \frac{P}{B\,W^{\frac{1}{2}}} \cdot f(\frac{a}{W})$$

$$f(\frac{a}{W}) = \frac{(2+\frac{a}{W})[0.886+4.64(\frac{a}{W})-13.32(\frac{a}{W})^2+14.72(\frac{a}{W})^3-5.6(\frac{a}{W})^4]}{(1-\frac{a}{W})^{\frac{3}{2}}}$$

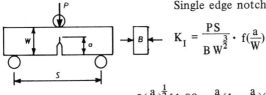

Single edge notched bend specimen (SENB):

$$K_I = \frac{PS}{B\,W^{\frac{3}{2}}} \cdot f(\frac{a}{W})$$

$$f(\frac{a}{W}) = \frac{3(\frac{a}{W})^{\frac{1}{2}}[1.99-\frac{a}{W}(1-\frac{a}{W})(2.15-3.93(\frac{a}{W})+2.7(\frac{a}{W})^2)]}{2(1+2\frac{a}{W})(1-\frac{a}{W})^{\frac{3}{2}}}$$

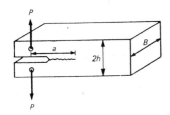

Double cantilever beam specimen (DCB):

$$K_I = 2\sqrt{3}\,\frac{Pa}{Bh^{\frac{3}{2}}} \qquad \text{(plane stress)}$$

$$K_I = \frac{2\sqrt{3}}{\sqrt{1-\nu^2}}\,\frac{Pa}{Bh^{\frac{3}{2}}} \quad \text{(plane strain)}$$

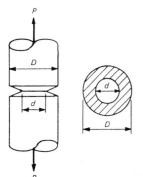

Circumferentially notched bar:

$$K_I = \frac{0.932\,P\sqrt{D}}{\sqrt{\pi d^2}}$$

over the range $1.2 \leqslant D/d \leqslant 2.1$

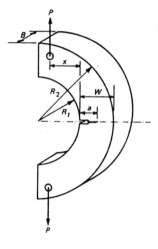

C-shaped specimen:

$$K_I = \frac{P}{BW^{\frac{1}{2}}} [1 + 1.54(\frac{x}{W}) + 0.50(\frac{a}{W})][1 + 0.221(1 - (\frac{a}{W})^{\frac{1}{2}})(1 - \frac{R_1}{R_2})] \cdot f(\frac{a}{W})$$

$$f(\frac{a}{W}) = 18.23(\frac{a}{W})^{\frac{1}{2}} - 106.2(\frac{a}{W})^{\frac{3}{2}} + 389.7(\frac{a}{W})^{\frac{5}{2}} - 582.0(\frac{a}{W})^{\frac{7}{2}} + 369.1(\frac{a}{W})^{\frac{9}{2}}$$

Useful Solutions for Practical Applications

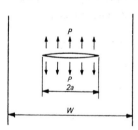

Crack under internal pressure:

$$K_I = CP\sqrt{\pi a} \quad \text{per unit thickness}$$

where C is the same as for the centre cracked plate.

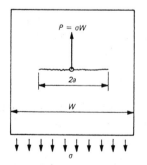

Cracks growing from both sides of a loaded hole, where the hole is small with respect to the crack:

$$K_I = C(\frac{\sigma\sqrt{\pi a}}{2} + \frac{P}{2\sqrt{\pi a}}) \quad \text{per unit thickness}$$

where C is the same as for the centre cracked plate.

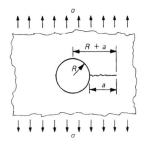

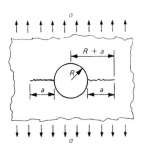

Through thickness cracks from one or both sides of a remotely loaded hole:

$$K_I = \sigma \sqrt{\pi a} \ F_1\left(\frac{a}{R+a}\right) \qquad \text{(single crack)}$$

$$K_I = \sigma \sqrt{\pi a} \ F_2\left(\frac{a}{R+a}\right) \qquad \text{(double crack)}$$

The solutions for F_1 and F_2 are given in figure 2.16.

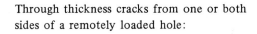

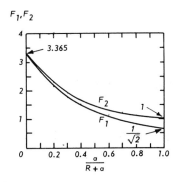

Figure 2.16. Correction factors for cracks growing from remotely loaded holes.

For long double cracks with $\dfrac{a}{R+a} > 0.3$ a good approximation is

$$K_I = C\sigma\sqrt{\pi(R+a)}$$

where C is the same as for the centre cracked plate.

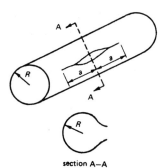

section A–A

Axial through thickness crack of length 2a in a thin walled pipe with mean radius R, internal pressure P and wall thickness t:

$$K = \sigma_H M_f \sqrt{\pi a}$$

where $\sigma_H = PR/t$ and M_f is the Folias correction factor for bulging of the crack flanks:

$$M_f = \sqrt{1 + 1.255\frac{a^2}{Rt} - 0.0135\frac{a^4}{R^2 t^2}}.$$

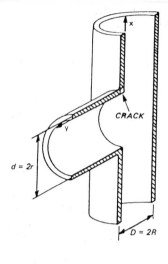

$d = 2r$

$D = 2R$

Corner crack in a longitudinal section of a pipe-vessel intersection in a pressure vessel:

$$\frac{K_I}{\sigma_H \sqrt{\pi a}} = F_m \left[1 + \left(\frac{rt}{RB} \right)^{\frac{1}{2}} \right]$$

where σ_H is the hoop stress in the vessel wall. The solution for F_m is given in figure 2.17.

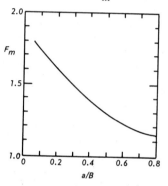

Figure 2.17. Correction factor for a corner crack in a longitudinal section of a pipe-vessel intersection in a pressure vessel.

DETAILED CRACK GEOMETRY

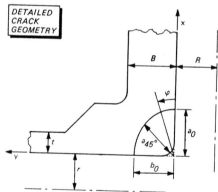

Note: This is a serious problem in pressure vessel technology. The solution given here is by M.A. Mohamed and J. Schroeder, International Journal of Fracture, Vol. 14, p. 605, 1978. It is a first approximation only, based on correlation of results from six different investigations. For detailed problem solving the individual solutions should be studied to ascertain their relevance.

2.9. Bibliography

1. Westergaard, H.M., *Bearing Pressures and Cracks*, Transactions of the ASME, Vol. 61, pp. A49–A53 (1939).
2. Paris, P.C. and Sih, G.C., *Stress Analysis of Cracks*, Fracture Toughness Testing and

its Applications, ASTM STP 381, pp. 63–77 (1965): Philadelphia.

3. Maddox, S.J., *An Analysis of Fatigue Cracks in Fillet Welded Joints,* International Journal of Fracture, Vol. 11, pp. 221–243 (1975).

4. Raju, I.S. and Newman, J.C. Jr., *Stress Intensity Factors for a Wide Range of Semi-Elliptical Surface Cracks in Finite Thickness Plates,* Engineering Fracture Mechanics, Vol. 11, pp. 817–829 (1979).

5. Newman, J.C. Jr. and Raju, I.S., *An Empirical Stress Intensity Factor Equation for the Surface Crack,* Engineering Fracture Mechanics, Vol. 15, pp. 185–192 (1981).

6. Tada, H., Paris, P.C. and Irwin, G.R., The Stress Analysis of Cracks Handbook, Del Research Corporation (1973): Hellertown, Pennsylvania.

7. Rooke, D.P. and Cartwright, D.J., Compendium of Stress Intensity Factors, Her Majesty's Stationery Office (1976): London.

8. Sih, G.C. (Editor), Methods of Analysis and Solutions of Crack Problems, Noordhoff International Publishing (1973): Leyden.

9. Cartwright, D.J. and Rooke, D.P., *Approximate Stress Intensity Factors Compounded from Known Solutions,* Engineering Fracture Mechanics, Vol. 6, pp. 563–571 (1974).

10. Paris, P.C., McMeeking, R.M. and Tada, H., *The Weight Function Method for Determining Stress Intensity Factors,* Cracks and Fracture, ASTM STP 601, pp. 471–489 (1976): Philadelphia.

11. Pickard, A.C., *Stress Intensity Factors for Cracks with Circular and Elliptic Crack Fronts, Determined by 3D Finite Element Methods,* Numerical Methods in Fracture Mechanics, Pineridge Press, pp. 599–614 (1980): Swansea.

12. Raju, I.S. and Newman, J.C. Jr., *Stress Intensity Factors for Two Symmetric Corner Cracks,* Fracture Mechanics, ASTM STP 677, pp. 411–430 (1979): Philadelphia.

3. CRACK TIP PLASTICITY

3.1. Introduction

In chapter 2 the elastic stress field equations for a sharp crack, equations (2.10), were obtained. These equations result in infinite stresses at the crack tip, i.e. there is a stress singularity. As has already been mentioned in section 1.7, structural materials deform plastically above the yield stress, and so in reality there will be a plastic zone surrounding the crack tip.

Along the x-axis $\theta = 0$ and the expression for σ_y in equations (2.10) gives

$$\sigma_y = \frac{\sigma\sqrt{\pi a}}{\sqrt{2\pi r}} = \frac{K_I}{\sqrt{2\pi r}}. \tag{3.1}$$

By substituting the yield strength, σ_{ys}, for σ_y in equation (3.1) an estimate can be obtained of the distance r_y over which the material is plastically deformed ahead of the crack:

$$r_y = \frac{1}{2\pi}\left(\frac{K_I}{\sigma_{ys}}\right)^2. \tag{3.2}$$

Assuming as a first approximation that r_y corresponds to the diameter of a circular plastic zone, the distribution of σ_y ahead of the crack tip will be as shown in figure 3.1. From this figure it is clear that the assumption is inaccurate, since part of the stress distribution is simply cut off above σ_{ys}. Also, there is no a priori reason why the plastic zone should be circular.

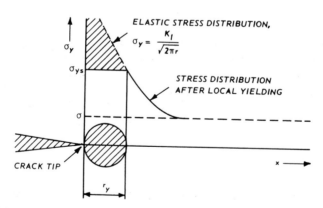

Figure 3.1. A first approximation to the crack tip plastic zone.

In fact it turns out to be extremely difficult to give a proper description of plastic zone size and shape. For this reason the models most widely known from the literature have followed one of two approaches. Either they give a better approximation of the size but use a selected shape, e.g. the Irwin and

Dugdale approaches discussed in sections 3.2 and 3.3, or they give an impression of the shape but retain the first size approximation, as in the derivations from classical yield criteria in section 3.4.

Besides these limitations there is the problem that the state of stress, i.e. plane stress or plane strain, will affect the plastic zone size and shape. It is well known that under plane strain conditions yielding need not occur until the applied stress is much higher than σ_{ys}, i.e. the plastic zone may be smaller. This and other effects of differing state of stress will be dealt with in sections 3.5 and 3.6.

Finally, in section 3.7 some remarks will be addressed to advanced methods of determining plastic zone size and shape.

3.2. The Plastic Zone Size According to Irwin

Irwin's analysis of plastic zone size attempts to account for the fact that the stress distribution cannot simply be cut off above σ_{ys} as in figure 3.1. For the analysis to be straightforward there are several restrictions:

1) The plastic zone shape is considered to be circular.
2) Only the situation along the x-axis ($\theta = 0$ in equations (2.10)) is analysed.
3) The material is considered to be elastic − perfectly plastic, i.e. stresses cannot exceed σ_{ys}.

Irwin argued that the occurrence of plasticity makes the crack behave as if it were longer than its physical size − the displacements are longer and the stiffness is lower than in the elastic case, i.e.

$$a_{eff} = a + \Delta a_n$$

where a_{eff} is the effective, or notional, crack length and Δa_n corresponds to the notional crack increment. Δa_n must account for redistribution of stresses that were above σ_{ys} in the elastic case, i.e. the stress distribution is not cut off as in the first approximation discussed in section 3.1. Since no strain hardening is allowed (restriction 3) Δa_n behaves as part of the crack.

Now consider the situation in figure 3.2. For a crack of length $a + \Delta a_n$

$$r_y = \frac{1}{2\pi}(\frac{K_I}{\sigma_{ys}})^2 = \frac{\sigma^2}{2\sigma_{ys}^2}(a + \Delta a_n). \tag{3.3}$$

For all stresses to be transmitted the area $\sigma_{ys} \cdot \Delta a_n$ should be equal to area A:

$$\sigma_{ys} \cdot \Delta a_n = \int_0^{r_y} \frac{\sigma\sqrt{\pi(a + \Delta a_n)}}{\sqrt{2\pi r}} dr - \sigma_{ys} \cdot r_y \tag{3.4}$$

$$\sigma_{ys}(\Delta a_n + r_y) = \int_0^{r_y} \frac{\sigma\sqrt{\pi(a + \Delta a_n)}}{\sqrt{2\pi}} \frac{dr}{\sqrt{r}}$$

$$= \frac{2\sigma\sqrt{a + \Delta a_n}}{\sqrt{2}} \sqrt{r_y}. \tag{3.5}$$

Using equation (3.3) we can substitute for $\sigma\sqrt{a + \Delta a_n}$ in equation (3.5). This gives

$$\sigma_{ys}(\Delta a_n + r_y) = \frac{2\sigma_{ys}\sqrt{2r_y}\sqrt{r_y}}{\sqrt{2}}$$

and therefore

$$\Delta a_n + r_y = 2r_y. \tag{3.6}$$

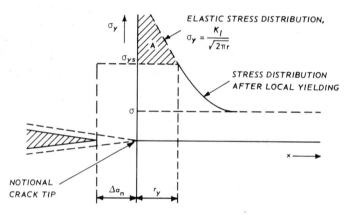

Figure 3.2. Schematic of Irwin's analysis.

Thus Irwin's analysis results in a plastic zone diameter twice that obtained as a first approximation. Furthermore, this result means that the notional crack length extends to the centre of the circular plastic zone, figure 3.3, with a concomitant shift of the stress distribution a distance r_y with respect to the elastic case. It is to be noted that the elastic stress distribution $\sigma_y = K_I/\sqrt{2\pi r}$ takes over from σ_{ys} at a distance $2r_y$ ahead of the actual crack tip. Since the same K always gives the same plastic zone size, equation (3.2), the stresses and strains both inside and outside the plastic zone will be determined by K. Hence the stress intensity approach is still applicable for correlating crack growth and fracture behaviour.

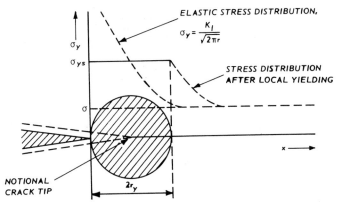

Figure 3.3. The Irwin plastic zone size.

A Useful Expression: Crack Tip Opening Displacement (COD)

In section 2.3 the crack flank displacement was defined. This is shown again in figure 3.4. The expression obtained was $v = 2\sigma\sqrt{a^2 - x^2}/E$. Clearly, the total crack opening, $\delta = 2v$, is $4\sigma\sqrt{a^2 - x^2}/E$. If crack tip plasticity is not accounted for, the displacement and crack opening, δ_t, at the crack tip are equal to zero. However, if Irwin's proposal to use an effective crack length is adopted then

$$\delta_t = \frac{4\sigma}{E} \sqrt{a^2 + 2ar_y + r_y^2 - a^2}$$

$$\sim \frac{4\sigma}{E} \sqrt{2ar_y}.$$

Substitution for r_y from equation (3.2) gives

$$\delta_t = \frac{4}{\pi} \frac{K_I^2}{E\sigma_{ys}} \tag{3.7}$$

which is an approximation for the Crack Tip Opening Displacement (COD). This approximation will be compared in section 3.3 with the more usual expressions for COD, which are derived from the Dugdale approach.

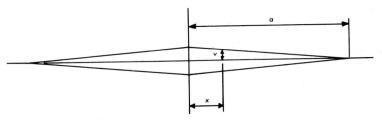

Figure 3.4. Crack flank displacement.

3.3. The Plastic Zone Size According to Dugdale

Dugdale's analysis assumes that all plastic deformation concentrates in a strip in front of the crack, the strip yield model. This type of behaviour does indeed occur for a number of materials, but certainly not for all. As in Irwin's analysis, Dugdale argued that the effective crack length is longer than the physical length. However, in this case the notional crack increment is considered to carry the yield stress as shown in figure 3.5 (the assumption of elastic-perfectly plastic behaviour is also made).

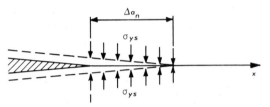

Figure 3.5. Schematic of Dugdale's analysis.

The Dugdale approach proceeds as follows:

1) Obtain a Westergaard-type stress function for a crack of length $2(a + \Delta a_n)$ with the origin at the centre of the crack. From section 2.2 a suitable function is known to be $\phi_1(z) = \sigma/\sqrt{1 - (a + \Delta a_n)^2/z^2}$, see equation (2.6).

2) Since Δa_n carries the yield stress, unlike a real crack, the elastic stress function $\phi_1(z)$ will overestimate the stress intensity at the notional crack tip. To obtain the correct estimate use is made of the superposition principle discussed in remark 3 appended to section 2.2. Thus a stress function that describes the loading condition over the distance Δa_n must be found, and this stress function must be subtracted from $\phi_1(z)$.

3) The loading condition for Δa_n can be described by the integral of a stress function $\phi_2(z)$ representing a point force σ_{ys} at a distance b from the actual crack tip. The solution for $\phi_2(z)$, given in reference 1 of the bibliography to this chapter, is

$$\phi_2(z) = \frac{2\sigma_{ys}z\sqrt{(a + \Delta a_n)^2 - b^2}}{\pi\sqrt{z^2 - (a + \Delta a_n)^2} \, (z^2 - b^2)} \tag{3.8}$$

and the required integral (also a stress function) is

$$\phi_3(z) = \int_a^{a+\Delta a_n} \frac{2\sigma_y z z}{\pi\sqrt{z^2 - (a + \Delta a_n)^2}} \cdot \frac{\sqrt{(a + \Delta a_n)^2 - b^2}}{z^2 - b^2} \cdot db$$

$$\phi_3(z) = \frac{2\sigma_{ys}}{\pi}\left[\frac{z}{\sqrt{z^2 - (a + \Delta a_n)^2}} \arccos(\frac{a}{a + \Delta a_n}) - \text{arccot}(\frac{a}{z}\sqrt{\frac{z^2 - (a + \Delta a_n)^2}{(a + \Delta a_n)^2 - a^2}})\right]$$

$$\tag{3.9}$$

4) The correct stress function, $\phi_4(z)$, is $\phi_1(z) - \phi_3(z)$, i.e.

$$\phi_4(z) = \frac{\sigma z}{\sqrt{z^2 - (a + \Delta a_n)^2}} - \frac{2\sigma_{ys}}{\pi} \frac{z}{\sqrt{z^2 - (a + \Delta a_n)^2}} \arccos(\frac{a}{a + \Delta a_n})$$

$$+ \frac{2\sigma_{ys}}{\pi} \operatorname{arccot}(\frac{a}{z} \sqrt{\frac{z^2 - (a + \Delta a_n)^2}{(a + \Delta a_n)^2 - a^2}}). \qquad (3.10)$$

5) Dugdale continued with the argument that a stress singularity cannot exist at the notional crack tip, since at that point the elastic stress goes no higher than the yield stress for an elastic-perfectly plastic material. This means that the singular terms in equation (3.10) must cancel each other, i.e.

$$\frac{\sigma z}{\sqrt{z^2 - (a + \Delta a_n)^2}} - \frac{2\sigma_{ys}}{\pi} \frac{z}{\sqrt{z^2 - (a + \Delta a_n)^2}} \arccos(\frac{a}{a + \Delta a_n}) = 0$$

so

$$\sigma - \frac{2\sigma_{ys}}{\pi} \arccos(\frac{a}{a + \Delta a_n}) = 0$$

and

$$\frac{\pi\sigma}{2\sigma_{ys}} = \arccos(\frac{a}{a + \Delta a_n}) \quad \text{or} \quad \cos(\frac{\pi\sigma}{2\sigma_{ys}}) = \frac{a}{a + \Delta a_n}. \qquad (3.11)$$

6) Finally the extent of the plastic zone, Δa_n, is found by expanding $\cos(\pi\sigma/2\sigma_{ys})$ as a series and approximating $a/(a + \Delta a_n)$ by $(a - \Delta a_n)/a$. The result is

$$\frac{a - \Delta a_n}{a} = 1 - \frac{\pi^2 \sigma^2}{8\sigma_{ys}^2}$$

and

$$\Delta a_n = \frac{\pi^2 \sigma^2 a}{8\sigma_{ys}^2} = \frac{\pi}{8} (\frac{K_I}{\sigma_{ys}})^2. \qquad (3.12)$$

Remarks

1) The Dugdale plastic zone size, equation (3.12), is

$$\Delta a_n = 0.393 (\frac{K_I}{\sigma_{ys}})^2.$$

This is somewhat larger than the diameter of the plastic zone according to Irwin. Irwin's analysis gives a plastic zone diameter $2r_y$, which from equation (3.2) is

$$2r_y = \frac{1}{\pi}(\frac{K_I}{\sigma_{ys}})^2 = 0.318\,(\frac{K_I}{\sigma_{ys}})^2.$$

2) The Dugdale approach has been given here in its most general form by using stress functions. Although this might appear unnecessarily complicated as compared to the more common treatment in terms of stress intensity factors, there is an important advantage. Stress intensity factors can be used only as approximations consisting of singular terms: thus in step 5) of the analysis, when all singular terms are required to cancel each other, the misleading impression is given that the stresses at the notional crack tip (the end of the plastic zone) should be zero. However, equation (3.10) shows that there is a non-singular term

$$\phi_5(z) = \frac{2\sigma_{ys}}{\pi} \operatorname{arccot}(\frac{a}{z})\,\sqrt{\frac{z^2 - (a + \Delta a_n)^2}{(a + \Delta a_n)^2 - a^2}}\,) \qquad (3.13)$$

and this gives the elastic stress distribution beyond the plastic zone. Note that equation (3.13) indeed predicts that for $z = a + \Delta a_n$ (the boundary of the plastic zone) and $y = 0$, $\sigma_y = \operatorname{Re}\phi_5(z) = \frac{2}{\pi}\sigma_{ys}\operatorname{arccot}0 = \sigma_{ys}$.

The Dugdale Approach and COD

An important aspect of the Dugdale approach in terms of stress functions is that it enables a basic expression for the COD to be calculated. The crack flank displacement, v, in the region between a and $a + \Delta a_n$ can be obtained by substituting the non-singular term in equation (3.10), i.e. $\phi_5(z)$ from equation (3.13), into the plane stress equivalent of equation (2.18.b). The plane stress relation for v is obtained by writing equation (2.18.b) in terms of μ, where $\mu = E/2(1 + v)$, and substituting $v/1 + v$ for v see equations (2.21.a) and (2.22.a)

$$Ev = 2\operatorname{Im}\overline{\phi}(z) - y(1 + v)\operatorname{Re}\phi(z).$$

Now $y = 0$, since the Dugdale plastic zone is a strip yield model along the x-axis. Thus

$$v = \frac{2\operatorname{Im}\overline{\phi}_5(z)}{E}. \qquad (3.14)$$

The solution of equation (3.14) is fairly difficult and beyond the scope of this course. A full treatment is given in reference 2 of the bibliography. For our purpose it is sufficient to note that an expression for the physical COD, δ_t, is obtained by solving equation (3.14) for v and allowing z to tend to the limit a. Then

$$2v_t = \delta_t = \frac{8\sigma_{ys}a}{\pi E}\,\ln\sec(\frac{\pi\sigma}{2\sigma_{ys}}). \qquad (3.15)$$

Equation (3.15) is the starting point for all COD considerations in the literature. Further detailed attention is given to the COD concept in chapter 6 of this course, but it is here informative to compare the results from Irwin's and Dugdale's analysis. The ln sec expression in equation (3.15) can be expanded as a series when σ/σ_{ys} is much less than unity, as is the case for LEFM conditions. Then

$$\delta_t = \frac{8\sigma_{ys}a}{\pi E} \left[\frac{1}{2} \left(\frac{\pi\sigma}{2\sigma_{ys}} \right)^2 + \frac{1}{12} \left(\frac{\pi\sigma}{2\sigma_{ys}} \right)^4 + \ldots \right].$$

Neglecting all terms in the series beyond the first,

$$\delta_t = \frac{\pi\sigma^2 a}{E\sigma_{ys}} = \frac{K_I^2}{E\sigma_{ys}}. \tag{3.16}$$

This value of COD is slightly less than that obtained via Irwin's analysis, equation (3.7):

$$\delta_t = \frac{4}{\pi} \frac{K_I^2}{E\sigma_{ys}} = 1.27 \frac{K_I^2}{E\sigma_{ys}}.$$

Note: Plane Stress and Plane Strain

So far, all expressions in Irwin's and Dugdale's analyses have been derived for the state of plane stress. Differences that arise owing to plane strain conditions will be discussed in section 3.5.

3.4. First Order Approximations of Plastic Zone Shapes

In the introduction to this chapter, section 3.1, it was mentioned that well known models describing the crack tip plastic zone fall into two categories. Either they estimate the size of a zone with an assumed shape, or else the shape is determined from a first order approximation to the size. Having dealt with the first category in sections 3.2 and 3.3, we shall now turn to methods for assessing the plastic zone shape.

The reason that a first order approximation to the plastic zone size is used in well known models for determining the shape is that the calculations employ classical yield criteria, e.g. those of Von Mises or Tresca, to give only the boundaries where the material starts to yield. No account is taken of the fact that the original elastic stress distribution above σ_{ys} must be redistributed and retransmitted. The procedure is similar to that in section 3.1, but instead of calculating r_y only for $\theta = 0$, the value of r_y over the range $(-\pi \leqslant \theta \leqslant +\pi)$ is determined.

Derivation of the plastic zone shape is thus simply a matter of substituting the appropriate stress equations into the yield criterion under consideration. Only the Von Mises criterion will be employed here, since the Tresca criterion gives similar results. However, both plane stress and plane strain plastic zone shapes will be analysed. The plane strain plastic zone shape is included in this section because the procedure for determining it is the same as for the plane stress case, and the result serves as a good basis for discussing the problem of differing states of stress in section 3.5.

Plastic Zone Shapes from the Von Mises Yield Criterion

The Von Mises yield criterion states that yielding will occur when

$$(\sigma_1 - \sigma_2)^2 + (\sigma_2 - \sigma_3)^2 + (\sigma_3 - \sigma_1)^2 = 2\sigma_{ys}^2 \qquad (3.17)$$

where σ_1, σ_2 and σ_3 are the principal stresses. In section 2.3 the mode I stress field equations were derived in terms of the principal stresses, namely

$$\sigma_1 = \frac{K_I}{\sqrt{2\pi r}} \cos\frac{\theta}{2} (1 + \sin\frac{\theta}{2})$$

$$\sigma_2 = \frac{K_I}{\sqrt{2\pi r}} \cos\frac{\theta}{2} (1 - \sin\frac{\theta}{2})$$

and σ_3 is either 0 (plane stress) or $\nu(\sigma_1 + \sigma_2)$ for plane strain. Substitution into equation (3.17) gives for plane stress

$$\frac{K_I^2}{2\pi r} (1 + \frac{3}{2}\sin^2\theta + \cos\theta) = 2\sigma_{ys}^2$$

or

$$r(\theta) = \frac{1}{4\pi} (\frac{K_I}{\sigma_{ys}})^2 (1 + \frac{3}{2}\sin^2\theta + \cos\theta). \qquad (3.18)$$

Equation (3.18) can be made dimensionless by dividing by r_y, the first order approximation to plastic zone size for $\theta = 0$, equation (3.2). Then

$$\frac{r(\theta)}{r_y} \text{ plane stress} = \frac{1}{2} + \frac{3}{4}\sin^2\theta + \frac{1}{2}\cos\theta. \qquad (3.19)$$

Note that for $\theta = 0$ the value of $r(\theta)$ is indeed r_y.

For plane strain, i.e. $\sigma_3 = \nu(\sigma_1 + \sigma_2)$

$$\frac{K_I^2}{2\pi r}\left[\frac{3}{2}\sin^2\theta + (1 - 2\nu)^2(1 + \cos\theta)\right] = 2\sigma_{ys}^2$$

and $\quad\dfrac{r(\theta)}{r_y}$ plane strain $= \dfrac{3}{4}\sin^2\theta + \dfrac{1}{2}(1 - 2\nu)^2(1 + \cos\theta).$ $\qquad$ (3.20)

Along the x-axis ($\theta = 0$) the plane strain value of $r(\theta)$ is much less than the plane stress value. Assuming $\nu = 0.33$,

$$r(\theta) \text{ plane strain} = 0.11\ r(\theta) \text{ plane stress} = 0.11\ r_y.$$

Figure 3.6 depicts the shapes of the plane stress and plane strain plastic zones in dimensionless form.

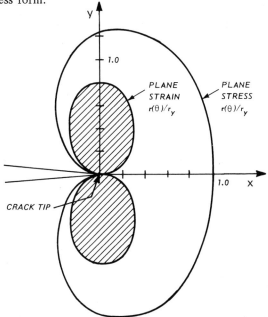

Figure 3.6. Dimensionless plastic zone shapes from the Von Mises yield criterion.

Similar derivations of plastic zone shapes can be obtained for mode II and mode III loading. Results of such derivations are given in reference 3 of the bibliography to this chapter.

3.5. The State of Stress in the Crack Tip Region

In section 3.1 it was mentioned that the state of stress, i.e. plane stress or plane strain, affects the plastic zone size and shape, and figure 3.6 is a good illustration of such effects. For this reason alone it is of interest to go into some detail concerning the state of stress in the crack tip region. However, there are additional important effects of stress state, and these will be discussed later in the present section and in section 3.6.

Through-Thickness Plastic Zone Size and Shape

Consider a through-thickness crack in a plate. From equations (2.10) in section 2.2 we know that there is at least a biaxial (plane stress) condition, for which the elastic stresses in the x and y directions are given by

$$\sigma_{ij} = \frac{\sigma\sqrt{\pi a}}{\sqrt{2\pi r}} \cdot f_{ij}(\theta). \tag{3.21}$$

Equation (3.21) shows that for small values of r both σ_x and σ_y will exceed the material yield stress. Thus a biaxial plastic zone will form at the crack tip. Assuming in the first instance that there is a uniform state of plane stress and that the plastic zone is circular as in Irwin's analysis, then a section through the plate in the plane of the crack gives the situation shown in figure 3.7. With no strain hardening the material within the plastic zone should be able to flow freely and contract in the thickness direction: however, the adjacent (and surrounding) elastic material cannot contract to the same extent. This leads to tensile stresses on the plastic zone boundary and in the thickness direction, i.e. a triaxial stress condition which when unrelieved by deformation would correspond to plane strain.

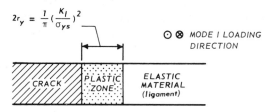

Figure 3.7. Schematic section in the crack plane.

In fact there is an interaction of stress states that can be described at best only semi-quantitatively. At the plate side surface there are no stresses in the thickness direction and so there is a biaxial condition of plane stress. Proceeding inwards there is an increasing degree of triaxiality that approaches and eventually may correspond to plane strain. Thus in a first approximation the plastic zone size and shape may be considered to vary through the thickness of the plate. For a plate of intermediate thickniss that is neither fully in plane stress nor predominantly in plane strain these approximate variations are considerable, as indicated schematically in figure 3.8. However, the plane stress surface regions will be more compliant than the plane strain interior, i.e. the surface regions will give a larger displacement v for the same remote stress σ, cf. equations (2.21) and (2.22). Consequently, load shedding occurs from the surface regions to the interior. This means that the plane stress and plane strain plastic zone sizes will be respectively smaller ald larger than those obtained from a first approximation. In practice, it has been found by finite element analysis that the through thickness plastic zone size variations are much less than those indicated schematically in figure 3.8, see reference 4 of the bibliography.

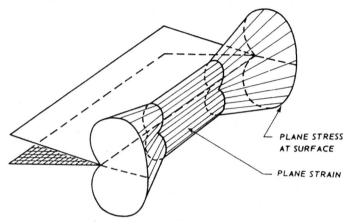

PLANE STRESS
AT SURFACE

PLANE STRAIN

Figure 3.8. Through-thickness plastic zone in a plate of intermediate thickness.

Calculation of the stress state distribution for a certain plate thickness is not possible. However, there are empirical rules for estimating whether the condition is predominantly plane stress or plane strain:

1) Full plane stress may be expected if the calculated size of the plane stress plastic zone, i.e. the diameter $2r_y$ in Irwin's analysis, is of the order of the plate thickness.

2) Predominantly plane strain may be expected when the calculated size of the plane stress plastic zone (which exists at the plate surfaces) is no larger than one-tenth of the plate thickness.

Through-Thickness Plastic Zone Size and the Plastic Constraint Factor

In chapter 2 the mode I stress field equations were expressed in terms of principal stresses, equations (2.13). From these expressions the relations between σ_1, σ_2 and σ_3 may be derived, namely

$$\sigma_2 = \sigma_1 \left(\frac{1 - \sin\frac{\theta}{2}}{1 + \sin\frac{\theta}{2}} \right)$$

and σ_3 is either 0 (plane stress) or $\nu(\sigma_1 + \sigma_2)$ for plane strain, i.e. $2\nu\sigma_1/(1 + \sin\frac{\theta}{2})$. It is readily seen that if $\theta = 0$, then $\sigma_2 = \sigma_1$ and $\sigma_3 = 2\nu\sigma_1$.

Assuming elastic conditions with $\nu = 0.33$ we can use the Von Mises yield criterion, equation (3.17), to determine the value of σ_1 that is reached before yielding occurs:

$$(\sigma_1 - 0.66\,\sigma_1)^2 + (0.66\,\sigma_1 - \sigma_1)^2 = 2\sigma_{ys}^2$$

and

$$\sigma_1 \approx 3\sigma_{ys}.$$

In this case it is said that the plastic constraint factor on yielding, C, is equal to 3.

If $C = 3$ then the first order approximation to the plane strain plastic zone size along the x-axis is

$$r(\theta) \text{ plane strain} = \frac{1}{2\pi} \left(\frac{K_I}{3\sigma_{ys}}\right)^2 = 0.11\ r_y.$$

This result was already obtained in section 3.4. The value of $r(\theta)$ plane strain must be a considerable underestimate of the overall through-thickness plastic zone size in a plate, since at the plate surfaces there is a state of plane stress and the plastic zone size will be r_y, i.e. nine times as large. For this reason Irwin proposed that an intermediate value of 1.68 be used for C, such that the nominal plane strain value of r_y is

$$\text{plane strain:} \quad r_y \approx \frac{1}{6\pi} \left(\frac{K_I}{\sigma_{ys}}\right)^2 \qquad (3.22)$$

Equation (3.22) is often quoted in the literature. However, in most work pertaining to the COD concept a value of $C = 2$ is used, see chapter 7.

Planes of Maximum Shear Stress

Besides plastic zone size and shape the state of stress also influences the locations of the planes of maximum shear stress in the vicinity of the crack tip. This is shown in figure 3.9 together with Mohr's circle construction for the principal stresses in plane stress and plane strain:

1) For $\theta = 0$ in plane stress the principal stresses σ_1 and σ_2 are σ_y and σ_x respectively, and $\sigma_3 = \sigma_z = 0$. This means that the maximum shear stress, τ_{max}, acts on $45°$ planes along the x-axis, figure 3.9.a.

---

*) Assuming the crack tips to have a finite radius is legitimate since a crack will always show some blunting due to plastic deformation. If the crack is considered to be slit-shaped with zero crack tip radius then equations (2.10) will apply and $\sigma_y = \sigma_1 = \sigma_x = \sigma_2 = \sigma_z = 0.5(\sigma_x + \sigma_y) = \sigma_3$. Therefore all principal stresses are equal ahead of a slit crack and there is a hydrostatic stress state in which τ_{max} is zero. Note that since stresses within the plastic zone are considered $\nu = 0.5$.

2) For plane strain the situation is slightly more complicated. We have to consider a crack with a finite tip radius*. In this case $\sigma_x\ (=\sigma_2)$ decreases to zero at the crack tip but $\sigma_y\ (=\sigma_1)$ does not, as can be seen by substituting $\theta = 0$ in equations (2.11) from section 2.2:

$$\sigma_x = \sigma_2 = \frac{K_I}{\sqrt{2\pi r}}\left(1 - \frac{\rho}{2r}\right) \qquad\qquad (3.23.a)$$

$$\sigma_y = \sigma_1 = \frac{K_I}{\sqrt{2\pi r}}\left(1 + \frac{\rho}{2r}\right) \qquad\qquad (3.23.b)$$

where $r = \rho/2$ at the crack tip. Thus σ_1 is always larger than σ_2 in the vicinity of the crack tip. Also, σ_3 changes from 0 (plane stress) to $\nu(\sigma_1 + \sigma_2)$. Because plastic deformation signifies that there is no change in volume ν must have a value of 0.5 within the plastic zone. Thus σ_3 is also larger than σ_2. Consequently the orientation of the planes of maximum shear stress changes to 45° along the z-axis, figure 3.9.b.

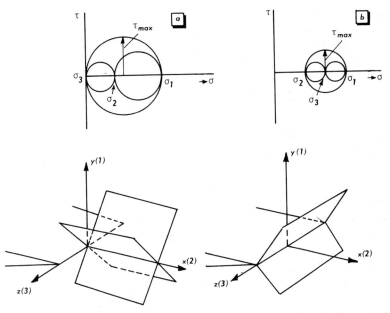

Figure 3.9. Location of the planes of maximum shear stress at the tip of a crack for a) plane stress and b) plane strain conditions.

Note that the situations depicted in figure 3.9 are valid only within the relatively small region of the plastic zone. The material will not shear off macroscopically along a plane of maximum shear stress, but will deform in a more complex manner as will be discussed in section 3.6.

3.6. Stress State Influences on Fracture Behaviour

In this section the effects of stress state on the macroscopic appearance of fracture and on the fracture toughness will be discussed.

Fracture Appearance

If a precracked specimen or component is monotonically loaded to fracture the general appearance corresponds to that sketched in figure 3.10. Crack extension begins macroscopically flat but is immediately accompanied by small 'shear lips' at the side surfaces. As the crack extends (which it does very quickly at instability) the shear lips widen to cover the entire fracture surface, which then becomes fully slanted either as single or double shear. This behaviour is usually attributed to crack extension under predominantly plane strain conditions being superseded by fracture under plane stress.

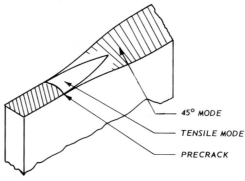

45° MODE

TENSILE MODE

PRECRACK

Figure 3.10. General appearance of monotonic overload fracture from a precrack.

An exact model for this flat-to-slant transition is not available, but it seems obvious that a change in the planes of maximum shear stress, see figure 3.9, plays an important role. Experimental studies by Hahn and Rosenfield (reference 5 of the bibliography to this chapter) indicate that under plane strain conditions a 'hinge' type deformation is followed by flat fracture, whereas under plane stress slant fracture occurs by shear from hinge type initiation. The two deformation modes are shown in figure 3.11.

The occurrence of slant fracture by shear is reasonably clear from figure 3.11.b. However, exactly how flat fracture results from hinge type deformation requires further explanation. According to Hahn and Rosenfield flat fracture occurs by tearing of material between the extensively deformed hinge type shear bands, figure 3.12. As to the possible causes of tearing the reader is referred to the last chapter of this course, section 13.4.

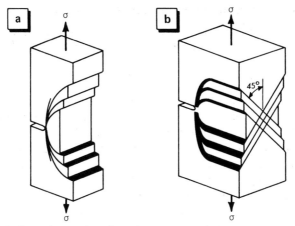

Figure 3.11. Deformation modes a) in plane strain and b) in plane stress.

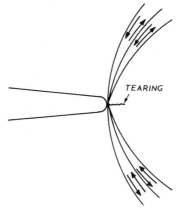

Figure 3.12. Tearing of material between hinge type shear bands.

Fracture Toughness

The critical stress intensity for fracture, K_c, depends on specimen thickness. A typical dependence is given in figure 3.13. Note that beyond a certain thickness, when the material is predominantly in plane strain and under maximum constraint, the value of K_c tends to a limiting constant value. This value is called the plane strain fracture toughness, K_{Ic}, and may be considered a material property.

The behaviour illustrated in figure 3.13 is generally ascribed to the plane stress − plane strain transition that occurs with increasing specimen thickness. However, a complete explanation for the observed effect of thickness does not exist. Also, the form of the K_c dependence for very thin specimens (less than 1 mm) is not exactly known, hence the dashed part of the plot. The most sa-

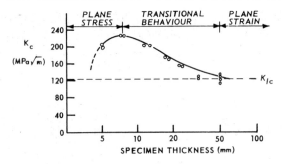

Figure 3.13. Variation in K_c with specimen thickness in a high strength maraging steel.

tisfactory model of the thickness effect is based on the energy balance approach and will be described in chapter 4.

Owing to the dependence of fracture toughness on specimen thickness and stress state it is evident that experimental determination of K_{Ic} will be possible only when specimens exceed a certain thickness. In turn this thickness will depend on the crack tip plastic zone size, as discussed in section 3.5, and therefore on the material yield strength. These and other aspects of plane strain fracture toughness determination and also the obtaining of K_c for thinner specimens will be dealt with in chapter 5, which concerns LEFM testing.

3.7. Some Additional Remarks on Plastic Zone Size and Shape Determination

In section 3.1 it was mentioned that it is extremely difficult to give a proper description of plastic zone size and shape. A detailed treatise on more advanced methods that attempt to do this is beyond the scope of this course, but an indication of the results is considered to be of interest here.

There are two general ways of tackling the problem: the experimental approach and finite element analysis using analytically determined boundary conditions.

The Experimental Approach

The most widely known work is that of Hahn and Rosenfield (references 5 and 6 of the bibliography to this chapter). They used specimens of silicon iron, which has the property that plastically deformed regions can be selectively etched and made visible. Some of the results are shown in figures 3.14 and 3.15. The specimen illustrated in figure 3.14 was in plane stress, and its plastic zone shape is schematically represented in figure 3.15. This shape is reasonably approximated by the Dugdale strip yield model. For plane strain the plastic zones were observed to closely resemble the shape in figure 3.6, which was derived from the Von Mises yield criterion.

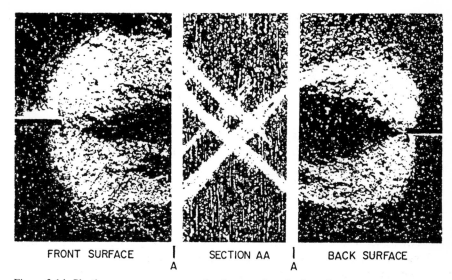

FRONT SURFACE | SECTION AA | BACK SURFACE
A A

Figure 3.14. Plastic zone appearance on the front surface, back surface and a normal section of a notched silicon iron specimen in plane stress, reference 6.

Figure 3.15. Schematic representation of the observations from figure 3.14.

Other experimental techniques include the use of electron microscopy, references 7 and 8 of the bibliography, and optical interferometry. These techniques are used mostly to study the development of plastic zones during fatigue crack growth in order to obtain more insight into the mechanisms of crack extension and also to check and refine crack growth models.

Finite Element Analysis

Of the various finite element analyses of plastic zone size and shape the work of Levy et al., reference 9 of the bibliography, is very well known. Figure 3.16 depicts their estimate for a plane strain crack tip plastic zone in an elastic -- perfectly plastic material together with the plastic zone derived in section 3.4 using the Von Mises yield criterion. It is seen that the latter, which is a first order approximation, is significantly smaller than the more accurate estimate provided by finite element analysis.

74

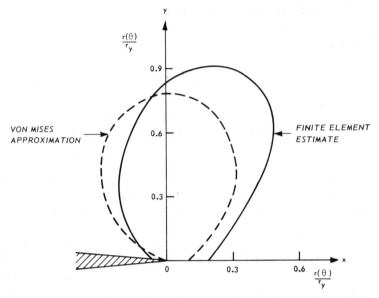

Figure 3.16. Comparison of plane strain plastic zone size and shape estimates for an elastic – perfectly plastic material.

3.8. Bibliography

1. Tada, H., Paris, P.C. and Irwin, G.R., The Stress Analysis of Cracks Handbook, Del Research Corporation (1973): Hellertown, Pennsylvania.
2. Burdekin, F.M. and Stone, D.E.W., *The Crack Opening Displacement Approach to Fracture Mechanics in Yielding*, Journal of Strain Analysis, Vol. 1, pp. 145–153 (1966).
3. McClintock, F.A. and Irwin, G.R., *Plasticity Aspects of Fracture Mechanics*, Fracture Toughness Testing and its Applications, ASTM STP 381, pp. 84–113 (1965): Philadelphia.
4. Prij, J., *Two and Three Dimensional Elasto-Plastic Finite Element Analysis of a Compact Tension Specimen*, Dutch Energy Research Centre Report ECN-80-178, Petten, The Netherlands, november 1980.
5. Hahn, G.T. and Rosenfield, A.R., *Local Yielding and Extension of a Crack under Plane Stress*, Acta Metallurgica, Vol. 13, pp. 293–306 (1965).
6. Rosenfield, A.R., Dai, P.K. and Hahn, G.T., *Crack Extension and Propagation under Plane Stress*, Proceedings of the First International Conference on Fracture, ed. T. Yokobori, T. Kawasaki and J.L. Swedlow, Japanese Society for Strength and Fracture of Materials, Vol. 1, pp. 223–258 (1966): Sendai, Japan.
7. Davidson, D.L. and Lankford, J., *Fatigue Crack Tip Plasticity Resulting from Load Interactions in an Aluminium Alloy*, Fatigue of Engineering Materials and Structures, Vol. 1, pp. 439–446 (1979).
8. Davidson, D.L. and Lankford, J., *The Influence of Water Vapour on Fatigue Crack Plasticity in Low Carbon Steel*, Proceedings of the First International Conference on Fracture, ed. D.M.R. Taplin, University of Waterloo Press, Vol. 2, pp. 897–904 (1977): Waterloo, Ontario.
9. Levy, N., Marcal, P.V., Ostergren, W.J. and Rice, J.R., *Small Scale Yielding near a Crack in Plane Strain: a Finite Element Analysis*, International Journal of Fracture Mechanics, Vol. 7, pp. 143–156 (1971).

4. THE ENERGY BALANCE APPROACH

4.1. Introduction

Besides the elastic stress field (stress intensity factor) approach there is another method that can be used in LEFM, the energy balance approach mentioned in sections 1.4 and 1.5 of chapter 1. In section 4.2 the equations of energy balance and instability will be given, together with the definition of the elastic energy release rate, G, which is the parameter controlling fracture.

In section 4.3 some important relations involving G are derived. One of these is the relationship between G and the change in compliance (inverse of stiffness) of a cracked specimen. This relationship has found much practical use and for that reason some applications of compliance determination are discussed in section 4.4.

In section 4.5 the energy balance interpretation of G is again considered in order to show its usefulness for materials exhibiting limited but significant plasticity. This leads to the concept of crack resistance, R, and the phenomenon of slow stable crack growth characterized by the R-curve, which is discussed in sections 4.6–4.8. The discussion of crack resistance is of a general nature and includes a possible explanation of R-curve shape and the effect of specimen thickness on fracture toughness.

The subject of R-curves is important, and extensive information on their determination and use is given in references 1 and 2 of the bibliography at the end of this chapter. Also, the determination of R-curves is included in chapter 5, which concerns LEFM testing.

4.2. The Griffith Energy Balance Approach

A simple energy balance derivation valid for constant displacement of the specimen edges has already been presented in section 1.4 of the introduction to this course. It was stated that the total energy content U of an elastic, remotely loaded cracked plate is

$$U = U_0 + U_a + U_\gamma - F, \tag{4.1}$$

where U_0 = elastic energy content of the loaded uncracked plate (a constant),

U_a = change in the elastic strain energy caused by introducing the crack in the plate,

U_γ = change in elastic surface energy caused by the formation of the crack surfaces,

F   = work performed by external forces.

Figure 4.1 shows a schematic plot of the variation of the total energy U as a function of crack length a.

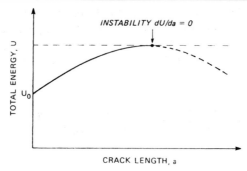

Figure 4.1. Schematic plot of the variation of total energy U as a function of crack length a.

Crack growth instability will occur as soon as U no longer increases with increasing crack length a.

Thus instability will occur if

$$\frac{dU}{da} \leqslant 0 \qquad (4.2.\text{a})$$

or since U_0 is a constant

$$\frac{d}{da}(U_a + U_\gamma - F) \leqslant 0. \qquad (4.2.\text{b})$$

Equation (4.2.b) can be rearranged to give:

$$\frac{d}{da}(F - U_a) \geqslant \frac{dU_\gamma}{da}. \qquad (4.3)$$

In the left hand part of equation (4.3) dF/da represents the energy provided by the external work F per unit crack extension and dU_a/da is the increase of elastic energy owing to the external work dF/da. Thus $dF/da - dU_a/da$ is the amount of energy that remains available for crack extension.

The right hand part, dU_γ/da, represents the elastic surface energy of the crack surfaces. This is the energy needed for the crack to grow, i.e. the crack resistance.

If we define the elastic energy release rate, G, per unit thickness as

$$G = \frac{d}{da}(F - U_a) \qquad (4.4)$$

and the crack resistance, R, per unit thickness as

$$R = \frac{d}{da}(U_\gamma), \qquad (4.5)$$

equation (4.3) can be rewritten as:

$$G \geqslant R. \qquad (4.6)$$

In section 1.4 the expression $U_a = \pi\sigma^2 a^2/E$ for the change in elastic strain energy caused by introducing a crack of length 2a in a plate of unit thickness was given without derivation. However, at this stage in the course it is possible to check the expression for U_a by considering crack flank displacements, figure 4.2.

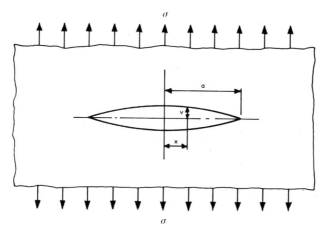

Figure 4.2. Crack flank displacement for a crack of length 2a in an infinite remotely loaded plate.

In section 2.3 the equations for crack flank displacements under plane strain and plane stress conditions were given, equations (2.21) and (2.22). For plane stress

$$v = \frac{2\sigma}{E} \sqrt{a^2 - x^2}. \qquad (4.7)$$

The elastic strain energy needed to open up the crack is $U_a = (P \times 2v)/2 = Pv$, where P is the force required for a total displacement of 2v. The force P may be obtained by summation of the stress σ over the total crack length, 2a. Thus

$$\text{plane stress} \quad U_a = 2 \int_0^a \sigma \cdot \frac{2\sigma}{E} \sqrt{a^2 - x^2} \; dx$$

$$= \frac{\pi\sigma^2 a^2}{E}. \qquad (4.8)$$

For plane strain the displacement v is obtained by multiplying the plane stress value by $(1 - \nu^2)$. Then

$$\text{plane strain} \quad U_a = (1 - \nu^2) \frac{\pi\sigma^2 a^2}{E}. \qquad (4.9)$$

Equations (4.8) and (4.9) apply to a situation where the load P remains constant when the crack is introduced. However, if the edges of the remotely loaded plate are fixed the introduction of the crack reduces the plate stiffness and the load will drop. In this case the change in elastic strain energy, U_a, is negative and equal to $-\pi\sigma^2 a^2/E$ for plane stress and $-(1 - \nu^2)(\pi\sigma^2 a^2/E)$ for plane strain.

The elastic surface energy, U_γ, is equal to the product of the elastic surface energy of the material, γ_e, and the surface area of the crack (two surfaces, length 2a). Therefore

$$U_\gamma = 2(2a\gamma_e). \tag{4.10}$$

Consequently, the instability condition for the case of plane stress (equation (4.3)) is

$$\frac{d}{da}(F - \frac{\pi\sigma^2 a^2}{E}) \geqslant \frac{d}{da}(4a\gamma_e). \tag{4.11}$$

For constant displacement of the specimen edges (the so called fixed grip condition) the external forces do not perform work during the crack extension process (F = constant). Thus $dF/da = 0$. Furthermore $U_a = -\pi\sigma^2 a^2/E$. Thus the instability condition for crack extension is

$$\frac{\pi\sigma^2 a}{E} \geqslant 2\gamma_e. \tag{4.12}$$

Fixed Grip and Constant Load Conditions

Equation (4.12) is valid only for the fixed grip condition. If instead there is a condition of constant load then crack extension results in increased displacement owing to decreased stiffness of the plate. The situation is thus more complicated than that for the fixed grip condition, but it can be analysed by assuming that it is possible to produce load-displacement diagrams for cracked plates as shown schematically in figure 4.3.

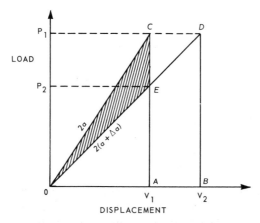

Figure 4.3. Assumed load-displacement diagram for a cracked specimen.

For both the fixed grip and constant load conditions the basic principle is the same, i.e. for crack extension to occur the elastic energy release should exceed the increase in surface energy. For the sake of clarity both the fixed grip and constant load conditions will be referred to in discussing figure 4.3. Under fixed grip conditions the load on a plate containing a crack of length $2a$ will drop from P_1 to P_2 when the crack extends at both sides by Δa and thereby reduces the stiffness of the cracked plate. This means that the elastic strain energy of the cracked plate, $U_0 + U_a$, decreases. Since the displacement of the ends of the plate does not alter, no external work is done. Thus the energy decrease is caused by U_a only (U_0 is always constant) and is represented by the shaded area OCE in figure 4.3.

For the constant load condition the same crack extension results in a displacement from v_1 to v_2. The load P_1 moves over this distance, and so the external force has done work equal to $P_1(v_2 - v_1)$. This work can be represented by the area ABCD in figure 4.3. On the other hand, the elastic strain energy U_a of the cracked plate has *increased* by an amount equal to area OBD − area OAC = $\frac{1}{2}P_1(v_2 - v_1)$ = area OCD. Thus from the amount of work $P_1(v_2 - v_1)$ done by the external forces only $P_1(v_2 - v_1) - \frac{1}{2}P_1(v_2 - v_1)$ = $\frac{1}{2}P_1(v_2 - v_1)$ = area OCD is available for crack extension, i.e. for constant load $F - U_a = U_a$. If we consider the first moment of crack extension, i.e. $\Delta a \to 0$, it is readily seen from figure 4.3 that area OCD → area OCE. Thus in the limit the *decrease* of elastic energy U_a of the cracked plate in the fixed grip condition (area OCE) is equal in magnitude to the *increase* of U_a in the constant load condition (area OCD). This implies that the elastic energy release rate G is equal in magnitude to dU_a/da, for the fixed grip condition as well as for the constant load condition. However, the sign is different for the two conditions: for fixed grips $G = -dU_a/da$ since U_a decreases; and for constant load $G = +dU_a/da$ since U_a is increased by the external work, which also provides the energy for crack extension.

Consider equation (4.4) once again, i.e.

$$G = \frac{d}{da}(F - U_a). \tag{4.4}$$

For the fixed grip condition F = constant and

$$G = -\frac{dU_a}{da}. \tag{4.13}$$

For constant load $F - U_a = U_a$ and consequently

$$G = \frac{d}{da}(F - U_a) = +\frac{dU_a}{da}. \tag{4.14}$$

The physical meaning of equation (4.14) is that for the constant load condition the work delivered by the external forces provides both the increase of elastic energy in the cracked plate and the energy necessary for crack extension.

4.3. Relations for Practical Use

Expressions for G

In the previous section it was shown that the elastic energy release rate, G, for infinitesimal crack extension is in magnitude equal to dU_a/da for constant load as well as for constant displacement. This being so, G may be obtained by differentiating equations (4.8) and (4.9) for a half crack length a:

plane stress $\qquad \dfrac{dU_a}{da} = G = \dfrac{\pi \sigma^2 a}{E}$

$$(4.15)$$

plane strain $\qquad \dfrac{dU_a}{da} = G = (1 - \nu^2)\,\dfrac{\pi \sigma^2 a}{E}.$

The Relation between G and K_I

A relation of prime importance is obtained by substituting $K_I = \sigma\sqrt{\pi a}$ in equations (4.15):

plane stress $\qquad G = \dfrac{K_I^2}{E}$

$$(4.16)$$

plane strain $\qquad G = \dfrac{K_I^2}{E}\,(1 - \nu^2)$

This direct relation between G and K_I means that under LEFM conditions the prediction of crack growth and fracture is the same for both the energy balance and elastic stress field approaches. As was mentioned in section 1.6, this equivalence was demonstrated by Irwin.

An Analytical Derivation of G

Although G has been indicated to be the controlling parameter for fracture according to the energy balance approach, a rigorous analysis has not yet been given. Such an analysis exists, however, notably that of Irwin (reference 3 of the bibliography to this chapter). We shall here present a more detailed version.

Consider a cracked body loaded by forces P as in figure 4.4. The general equation for the energy release rate, G, per unit thickness is:

$$G = \frac{d}{da}(F - U_a) \tag{4.4}$$

or

$$G = \frac{1}{B}\left(P\frac{dv}{da} - \frac{dU_a}{da}\right) = \frac{1}{B}\left(P\frac{dv}{da} - \frac{1}{2}\frac{dPv}{da}\right). \tag{4.17}$$

Note that the change in F is equal to the external work done, Pv, while the change in U_a is Pv/2. Introducing the compliance of the body, C, which is the inverse of its stiffness, i.e. $C = v/P$, equation (4.17) becomes

$$G = \frac{1}{B}\left(P\frac{dCP}{da} - \frac{1}{2}\frac{dCP^2}{da}\right)$$

$$G = \frac{1}{B} (PC \frac{dP}{da} + P^2 \frac{dC}{da} - \frac{1}{2} P^2 \frac{dC}{da} - CP \frac{dP}{da})$$

$$= \frac{P^2}{2B} \frac{dC}{da}. \qquad (4.18)$$

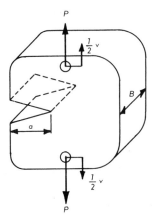

Figure 4.4. A cracked body loaded by forces P and with total displacement v
at the points of load application.

Now consider only the change of elastic energy, dU_a/da, of the cracked
body. We may write

$$\frac{1}{B} (\frac{dU_a}{da}) = \frac{1}{2B} (\frac{dPv}{da}) = \frac{1}{2B} (P \frac{dv}{da} + v \frac{dP}{da}). \qquad (4.19)$$

In the general case where neither P or v is kept constant, we obtain

$$\frac{1}{B} (\frac{dU_a}{da}) = \frac{1}{2B} (\frac{PdCP}{da} + \frac{CPdP}{da})$$

$$= \frac{1}{2B} (P^2 |\frac{dC}{da} + \frac{CPdP}{da} + \frac{CPdP}{da})$$

$$= \frac{P^2}{2B} \frac{dC}{da} + \frac{v}{B} \frac{dP}{da}. \qquad (4.20)$$

For the extreme case where load is kept constant, P = constant and consequently dP/da = 0. Thus equation (4.20) becomes

$$\frac{1}{B} (\frac{\partial U_a}{\partial a})_P = \frac{P^2}{2B} \frac{dC}{da}. \qquad (4.21)$$

For the other extreme case where displacement is kept constant, v = constant
and equation (4.20) becomes

$$\frac{1}{B} (\frac{\partial U_a}{\partial a})_v = \frac{P^2}{2B} \frac{dC}{da} + \frac{v}{B} \frac{d\frac{v}{C}}{da}$$

$$\frac{1}{B}\left(\frac{\partial U_a}{\partial a}\right)_v = \frac{P^2}{2B}\frac{dC}{da} + \frac{v}{B}\left(\frac{vdC^{-1}}{da} + \frac{1}{C}\frac{dv}{da}\right)_v$$

$$= \frac{P^2}{2B}\frac{dC}{da} + \frac{v^2}{B}\left(\frac{dC^{-1}}{dC}\frac{dC}{da}\right)$$

$$= \frac{P^2}{2B}\frac{dC}{da} - \frac{P^2}{B}\frac{dC}{da}$$

$$= -\frac{P^2}{2B}\frac{dC}{da}. \qquad (4.22)$$

From equations (4.18), (4.21) and (4.22) it follows that for fixed grips as well as for constant load G is equal in magnitude to $(1/B)(dU_a/da)$,

$$G = \frac{P^2}{2B}\frac{dC}{da} = \frac{1}{B}\left(\frac{\partial U_a}{\partial a}\right)_P = -\frac{1}{B}\left(\frac{\partial U_a}{\partial a}\right)_v. \qquad (4.23)$$

Equation 4.20 shows that if the load drops on crack extension (i.e. dP/da is negative), then dU_a/da decreases. This is the reason why there is a change of sign in going from the one extreme of constant load to the other extreme of constant displacement. Compare with the description in section 4.2.

Equation (4.23) also gives the explicit relation between G and the compliance, C. This important relation is the basis, together with equations (4.16), for using the compliance to determine stress intensity factors for certain specimen and crack geometries. The compliance technique is an addition to those methods discussed and listed in chapter 2, section 2.7, and will be illustrated in the next section.

4.4. Determination of Stress Intensity Factors from Compliance

From equations (4.16) and (4.23) it follows that

$$K_I^2 = E'G = \frac{E'P^2}{2B}\frac{dC}{da} \qquad (4.24)$$

where $E' = E$ for plane stress and $E/(1 - \nu^2)$ for plane strain. This general relation enables use of the compliance to determine stress intensity factors for certain specimen and crack geometries. A well known example is the double cantilever beam specimen (DCB) already mentioned in section 2.8 and depicted again in figure 4.5. From simple bending theory (neglecting shear displacements) the displacement v in the load line of the DCB specimen is given by

$$v = \frac{2Pa^3}{3EI} = \frac{8Pa^3}{EBh^3}.$$

Since $C = v/P$,

$$C = \frac{8a^3}{EBh^3} \quad \text{and} \quad \frac{dC}{da} = \frac{24a^2}{EBh^3}.$$

From equation (4.24)

$$K_I^2 = E'G = E' \frac{P^2}{2B} \frac{dC}{da} = \frac{E'}{E} \frac{12P^2 a^2}{B^2 h^3}$$

and

plane stress $\quad K_I = 2\sqrt{3} \; \dfrac{Pa}{Bh^{3/2}}$

$$(4.25)$$

plane strain $\quad K_I = \dfrac{2\sqrt{3}}{\sqrt{1-\nu^2}} \dfrac{Pa}{Bh^{3/2}}$

These expressions were given in section 2.8. Note that under elastic conditions ($\nu \sim 0.33$) the factor $1/\sqrt{1-\nu^2}$ gives a K_I value in plane strain only 6% larger than the plane stress value.

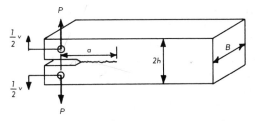

Figure 4.5. The double cantilever beam specimen (DCB).

A Constant K_I Specimen: The Tapered DCB

An interesting application of equation (4.25) is that a constant stress intensity factor may be obtained by keeping a/B or $a/h^{3/2}$ constant. The first possibility results in tapered thickness, which is not very useful since there will be a gradual change in stress state until full plane strain is reached. However, keeping $a/h^{3/2}$ constant or, still better, accounting for shear displacements by keeping $(3a^2 + h^2)/h^3$ constant, results in a constant K_I specimen, the tapered double cantilever beam (TDCB). This specimen, which is shown in figure 4.6, has found widespread use particularly in stress corrosion testing, and will be mentioned again in chapter 10. A detailed discussion of the TDCB specimen is given in reference 4 of the bibliography to the present chapter.

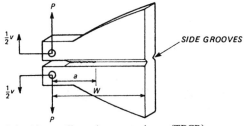

Figure 4.6. The tapered double cantilever beam specimen (TDCB).

Experimental Determination of C and K_I

Values of C and dC/da may be obtained experimentally as well as from theory. Load-displacement diagrams are made for a particular type of specimen containing cracks of different lengths, figure 4.7.a. The cracks may be extended artificially, e.g. a sawcut, or by fatigue loading. The compliance for different crack lengths is simply v/P, so that it is possible to construct a compliance versus crack length calibration diagram as in figure 4.7.b. From this diagram dC/da may be obtained and K_I may be calculated via equation (4.24), i.e.

$$K_I^2 = \frac{E'P^2}{2B} \frac{dC}{da}.$$

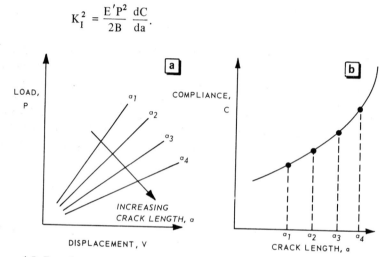

Figure 4.7. Experimental determination of compliance.

Note: Measurement of Displacements

For beam-type loaded specimens the crack flank displacements in the load line may be used instead of the exact displacements of the points of load application (shown schematically in figures 4.4 and 4.5). This is an important practical advantage since crack flank displacements are usually much easier to measure.

4.5. The Energy Balance for More Ductile Materials

The expressions and methods dealt with in sections 4.3 and 4.4 are all based on G being the controlling parameter for predicting fracture. However, G represents only the left-hand side of equation (4.3). The right-hand side of this equation, dU_γ/da, represents the surface energy increase owing to infinitesimal crack extension and is thus in the first instance to be considered as identical to the crack resistance, R, already mentioned in section 4.2. However, R is equal to the surface energy increase rate only for ideally brittle materials like glasses, ceramics, rocks and ice, which obey the original Griffith criterion. Reformulated in the manner suggested by Irwin (see also section

1.5) this criterion is:

$$G = \frac{\pi \sigma^2 a}{E} = G_c = 2\gamma_e = R = \text{a constant.} \qquad (4.26)$$

In 1948 Irwin and Orowan independently pointed out that the Griffith theory could be modified and applied to both brittle materials and metals that exhibit plastic deformation. The modification recognized that R is equal to the sum of the elastic surface energy, γ_e, and the plastic strain work, γ_p, accompanying crack extension. Consequently, equation (4.26) was changed to

$$G = \frac{\pi \sigma^2 a}{E} = G_c = 2(\gamma_e + \gamma_p) = R. \qquad (4.27)$$

For relatively ductile materials $\gamma_p \gg \gamma_e$, i.e. R is mainly plastic energy and the surface energy can be neglected. Also, it is no longer certain that instability and fracture will occur at a constant value of G_c, since R need not be a constant. In fact R, and hence G_c, are constant only for the condition of plane strain. In this case it is customary to write $R = G_{Ic}$, in an analogous way to the plane strain fracture toughness K_{Ic} discussed in section 3.6. Then the equilibrium condition for crack extension, i.e. instability, is given by equation (4.15):

plane strain $\quad G_{Ic} = (1 - \nu^2) \dfrac{\pi \sigma^2 a}{E}$

or $\sigma_c = \sqrt{\dfrac{EG_{Ic}}{(1 - \nu^2)\pi a}}.$ $\qquad (4.28)$

This condition can be represented graphically as in figure 4.8. It is clear from the foregoing relations that higher stresses result in instability at shorter crack lengths. (The dashed parts of the G lines in figure 4.8 are imaginary, since a certain crack length will be present at the beginning of loading.)

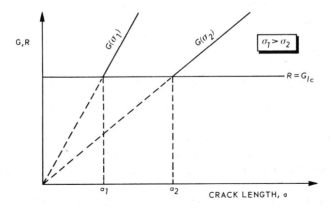

Figure 4.8. Graphical representation of the instability condition for plane strain.

A similar representation, figure 4.9, shows that beyond instability the equivalence of the fixed grip and constant load conditions no longer holds. For constant load G increases linearly with crack length (σ remains constant) but for fixed grips G increases less strongly, since the load and stress decrease as the crack extends and lowers the specimen stiffness, as was illustrated in figure 4.3. (Note that $\partial G/\partial a$ remains positive. This is because the specimen geometry, a remotely loaded centre cracked plate, results in dK_I/da and hence $\partial G/\partial a$ being positive. In chapter 5, section 5.4, the usefulness of specimens giving negative $\partial G/\partial a$ will be discussed.)

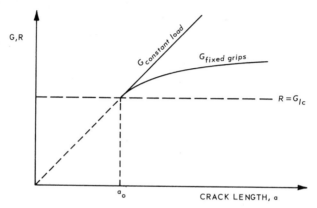

Figure 4.9. The difference in G for the fixed grip and constant load conditions beyond instability.

4.6. Slow Stable Crack Growth and the R-Curve Concept

As was mentioned in section 4.5, a constant value of R is obtained only for the condition of plane strain. For plane stress and intermediate plane stress — plane strain conditions it turns out that R is no longer constant. Loading a relatively thin (predominantly in plane stress) specimen containing a crack results in the behaviour shown schematically in figure 4.10. The

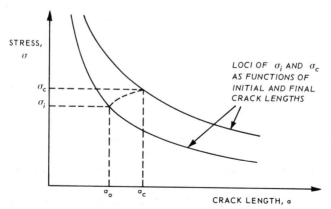

Figure 4.10. Slow stable crack growth in plane stress.

initial crack of length a_0 begins to extend at a certain stress σ_i. However, if the stress is maintained at σ_i no further crack growth occurs. A slight increase in the stress then results in additional crack extension, but the situation remains stable. The process of increasing stress accompanied by stable crack growth continues until a critical combination of stress, σ_c, and crack length, a_c, is reached, at which point instability occurs.

Instability is thus preceded by a certain amount of slow stable crack growth in specimens under full or predominantly plane stress. In terms of the energy balance approach this situation can be described as in figure 4.11. The value of R is depicted as a rising curve with a vertical segment corresponding to no crack extension at low stress (and G) levels. At a stress σ_i crack extension begins but R remains equal to G since the situation is stable. This is indicated by the fact that G_{σ_i} intersects the R-curve (G = R) but further crack growth cannot occur at σ_i because G_{σ_i} then becomes less than R. The stable condition is maintained until σ_c and a_c are reached. Beyond this point G > R, as indicated by the G_{σ_c} line, and instability occurs.

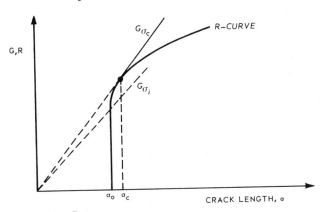

Figure 4.11. The rising R-curve.

Figure 4.11 shows that for instability to occur in plane stress it is not only necessary that G = R but also that the tangency condition $\partial G/\partial a = \partial R/\partial a$ be fulfilled. This second condition is a consequence of assuming a rising R-curve. This assumption has been verified experimentally, but as yet there is no definitive explanation as to why the R-curve rises: a possible explanation will be discussed in section 4.7. Meanwhile another important property of R-curves, their invariance with respect to initial crack length a_0, will be dealt with here.

R-Curve Invariance with Initial Crack Length

In Irwin's analysis R was considered to be independent of the total crack length a (= a_0 + Δa). This is true only for the plane strain condition, as illustrated in figure 4.8. If we now consider the crack resistance for a thin sheet, it is reasonable to assume that the very beginning of slow stable crack growth occurs under plane strain in the middle of the specimen thickness and is there-

88

fore independent of crack length. In addition, many tests have shown that the form of the rising part of the R-curve is also independent of crack length. Thus we may expect R-curves to be independent of the initial crack length a_0, i.e. an invariant R-curve may be placed anywhere along the horizontal axis of a (G,R)-crack length diagram, as in figure 4.12.

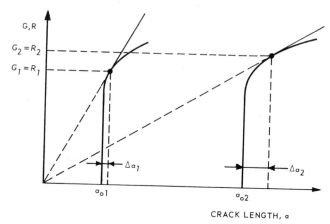

Figure 4.12. Invariant R-curves and the points of instability.

Figure 4.12 shows, however, that a shift of the complete R-curve along the crack length axis changes the point at which instability occurs, i.e. the tangency at which G = R and $\partial G/\partial a = \partial R/\partial a$. Hence the point of instability will depend on the initial crack length a_0. In summary, an invariant R-curve has the following consequences:

1) Crack initiation is independent of initial crack length a_0.
2) Instability depends on a_0. A longer a_0 results in more stable crack growth and a higher G value at instability. In comparison with the plane strain situation it may thus be stated that instability definitely depends on total crack length a $(= a_0 + \Delta a)$.

4.7. A Possible Explanation for the Rising R-Curve

Although no definitive analysis of the rising R-curve exists, a working hypothesis has been given by Krafft, Sullivan and Boyle (reference 5 of the bibliography). This hypothesis models the R-curve behaviour under intermediate plane stress – plane strain conditions.

Krafft et al. presuppose that in plane strain the plastic deformation energy necessary for crack extension is related to the area of newly created crack surface, but in plane stress the plastic energy is related to the volume contained by plane stress (45°) crack surfaces and their mirror images, as shown in figure 4.13. In this figure a specimen of thickness B is considered to crack with a fraction S of the thickness in plane stress and the remaining fraction 1 – S in plane strain. For a crack growth increment da the total energy consumption is

$$dW = (\frac{dW_S}{dA})\, B(1 - S)\, da + (\frac{dW_p}{dV}).\frac{B^2 S^2}{2}\, da \qquad (4.29)$$

where $\dfrac{dW_S}{dA}$ = energy consumption rate for plane strain (per unit surface area),

$\dfrac{dW_p}{dV}$ = energy consumption rate for plane stress (per unit volume).

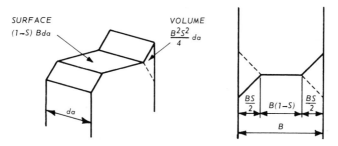

Figure 4.13. Model of Krafft et al. for explaining the development of crack resistance.

The crack resistance, R, is given by

$$R = \frac{1}{B}\frac{dW}{da} = \frac{dW_S}{dA}(1 - S) + \frac{dW_p}{dV}\frac{BS^2}{2}. \qquad (4.30)$$

Experimentally it has been found that $dW_p/dV \gg dW_S/dA$, so that from equation (4.30) it is evident that as soon as shear lips (slant fracture) start to form the value of R will show a sharp increase. In section 3.6 it was shown that monotonically loading to fracture generally results in a change from flat to slant fracture. This being so, the generality of the rising R-curve is confirmed.

An Explanation for the Effect of Specimen Thickness on Fracture Toughness

The effect of specimen thickness on fracture toughness was mentioned in section 3.6 and illustrated by figure 3.13. The model of Krafft et al. provides an explanation of this effect (for all but the thinnest specimens) in terms of the influence of specimen thickness on G_c, which is related to the critical stress intensity for fracture via equation (4.16), i.e. $K_c = \sqrt{E'G_c}$.

Knott (reference 6 of the bibliography) has made the reasonable assumption that the absolute thickness of shear lips is approximately constant, i.e. the fraction S in equation (4.30) decreases with increasing specimen thickness. On this basis Knott analysed data of Krafft et al. for the aluminium alloy 7075−T6, assuming a constant shear lip thickness of 2 mm. For a specimen 2 mm thick, i.e. $S = 1$, G_c had been found to be $200\,kJ/m^2$. In terms of equation (4.30) this means that

$$R = G_c = \frac{dW_p}{dV}\frac{B}{2} = 200\,kJ/m^2.$$

On the other hand a value of $20\,kJ/m^2$ was estimated for dW_S/dA. Thus equation (4.30) could be reduced to

$$G_c = 20\,(1 - S) + 200\,S^2.$$

Figure 4.14 shows the result of using this equation as compared to experimental data for the 7075-T6 alloy. The agreement is excellent and is strong evidence for the validity of the model of Krafft et al. In any event this model is the best currently available.

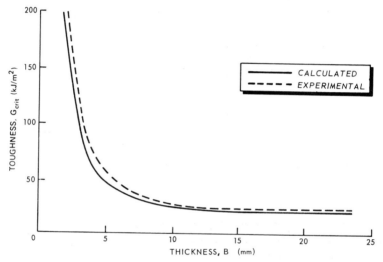

Figure 4.14. Knott's analysis of the model of Krafft et al. for the aluminium alloy 7075–T6.

4.8. Crack Resistance: a Complete Description

In sections 4.5 – 4.7 the concept of crack resistance, R, for more ductile materials has been developed. It was indicated that under plane strain conditions fracture takes place at a constant value of R and $G_c = G_{Ic}$, irrespective of the crack length. However, for plane stress and intermediate plane stress – plane strain conditions a rising R-curve develops and G_c is no longer constant and independent of crack length. Finally it was shown that the R-curve behaviour can be fairly well explained by the model of Krafft et al.

At this point it is desirable to give a complete description of the R-curve concept, but first it is necessary to discuss a change in notation. In the literature and in practice the R-curve is no longer considered in terms of G and R. Instead the stress intensity factors K_G and K_R are used. This is because the stress intensity factor concept has found widespread application and the energy balance parameters G and R may be simply converted to stress intensities via the relation $K_I = \sqrt{E'G}$.

The reason for not adopting this notational conversion in previous sections of this chapter is that the R-curve concept is based on energy balance principles and can be explained best in those terms. However, now it is convenient to describe the crack resistance behaviour in terms compatible with current practice.

The R-Curve in Terms of Stress Intensity Factors

A schematic R-curve in terms of K_G and K_R is presented in figure 4.15.

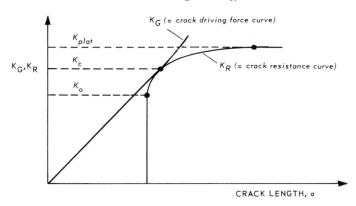

Figure 4.15. The R-curve in terms of stress intensity factor notation.

In this diagram there are three important points:
1) K_0 is the point of initial crack extension.
2) K_c is the critical stress intensity (instability point).
3) K_{plat} is the plateau level of the K_R-curve. (Note that this part of the curve is beyond instability.)

K_0 has been found by means of complicated experimental techniques to be independent of specimen thickness and to have a constant value for a particular material. The reason for this behaviour may be that even in relatively thin specimens the initial crack extension takes place in plane strain, as was assumed in section 4.6.

For K_c, however, there is a strong effect of specimen thickness: thinner specimens give higher K_c values and consequently exhibit more slow stable crack growth, since K_0 remains constant. A possible explanation for this behaviour was given in section 4.7. It should be noted that a sufficiently thick specimen will result in full plane strain and K_c will then be equal to K_0.

K_{plat} also depends strongly on specimen thickness. This parameter is not a generally accepted feature of the K_R-curve. A number of authors consider the existence of K_{plat} to be due to specimen finite geometry effects, and that the K_R-curve for very wide panels would attain a constant non-zero slope rather than a plateau level.

Figure 4.16 shows an example of experimentally determined values of K_0, K_c and K_{plat} as functions of specimen thickness. The data for K_{plat} (beyond instability) were obtained by a special testing technique to be described in the next chapter, section 5.4.

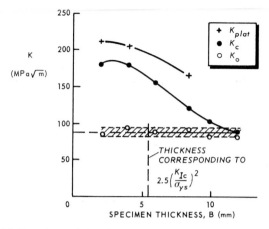

Figure 4.16. Experimental data for K_0, K_c and K_{plat} from reference 1 of the bibliography to this chapter.

Effect of Specimen Thickness on K_R-Curve Shape

From the observed dependence of K_0, K_c and K_{plat} on specimen thickness it is possible to indicate the general shapes of K_R-curves as functions of specimen thickness, figure 4.17. This figure demonstrates that a family of K_R-curves can be presented and that the curve for plane strain need not be considered as a separate case: it is simply a curve which does not show stable crack growth. In this respect the plane strain K_R-curve is analogous to the original (G,R) representation of instability, figure 4.8, since it is to be remembered that R-curves (and hence K_R-curves) are independent of the initial crack length, a_0. This analogy is schematically depicted in figure 4.18.

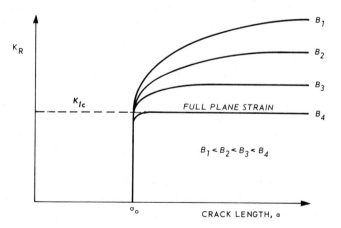

Figure 4.17. K_R-curves as functions of specimen thickness, B.

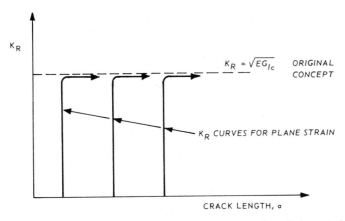

Figure 4.18. Comparison of the original and current representation of plane strain crack resistance.

Crack Resistance and K_{Ic} Testing

Although the procedure for K_{Ic} testing will be fully discussed in chapter 5, some remarks about the standard test method are appropriate here. This is because crack resistance test results, like those in figure 4.16, shed some light on problems in determining K_{Ic}.

It has been remarked several times that there is a thickness effect on fracture toughness, and that only above a certain specimen thickness are more or less constant values (i.e. K_{Ic}) obtained. The minimum required thickness for K_{Ic} determination has been found experimentally to be $2.5(K_{Ic}/\sigma_{ys})^2$. This thickness is indicated as a line in figure 4.16. It is seen that although K_0 is constant, K_c is still decreasing out to greater thickness.

Now the standard test method specifies determining K_{Ic} either from 'pop-in', which is an initial amount of audible stable crack growth, or from 2% crack extension. In view of the constancy of K_0 and the variation of K_c with specimen thickness, the question arises whether it is not so much the initial crack extension causing inconsistent K_{Ic} values, but the way in which this crack extension is defined: if it is stable a 2% crack extension can cause a significant increase in stress intensity. Furthermore, assuming that the representation of K_R-curves as in figure 4.17 is correct, then stable crack growth should have a small but consistent effect on the value of K_{Ic} when specimens of widely differing initial crack length are used. This effect is demonstrated schematically in figure 4.19 and has indeed been observed.

It might be thought that the foregoing problems could readily be avoided by using K_0 instead of K_{Ic}, but as mentioned earlier K_0 has to be found by means of complicated experimental techniques. This makes its determination unsuitable for a standard test.

94

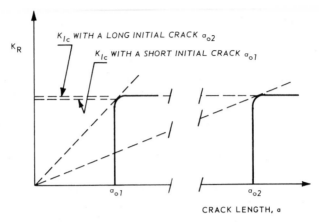

Figure 4.19. Effect of initial crack length a_0 on K_{Ic}.

4.9. Bibliography

1. Schwalbe, K.-H., *Mechanik und Mechanismus des Stabilen Risswachstums in Metallischen Werkstoffen*, Fortschrittsberichten der VDI Zeitschriften Reihe 18, No. 3 (1977).
2. McCabe, D.E. (Editor), Fracture Toughness Evaluation by R-Curve Methods, ASTM STP 527 (1973): Philadelphia.
3. Irwin, G.R., *Fracture*, Encyclopaedia of Physics, ed. S. Flüge, Springer Verlag, pp. 551–589 (1958): Berlin.
4. Mostovoy, S., Crosley, P.B. and Ripling, E.J., *Use of Crack-Line Loaded Specimens for Measuring Plane-Strain Fracture Toughness*, Journal of Materials, Vol. 2, pp. 661–681 (1967).
5. Krafft, J.M., Sullivan, A.M. and Boyle, R.W., *Effect of Dimensions on Fast Fracture Instability of Notched Sheets*, Proceedings of the Crack Propagation Symposium Cranfield, The College of Aeronautics, Vol. 1, pp. 8–28 (1962): Cranfield, England.
6. Knott, J.F., Fundamentals of Fracture Mechanics, Butterworths (1973): London.

5. LEFM TESTING

5.1. Introduction

In chapter 2 a number of analytical relationships were given for the determination of stress intensity factors in elastic specimens with different crack geometries. These stress intensity factors (K_I, K_{II} or K_{III} for the opening, edge-sliding or tearing modes of crack extension, figure 2.1) are functions of load, crack size and specimen geometry.

Ideally a critical stress intensity factor, K_c, associated with a specific crack geometry can be used to predict the behaviour in an actual structure. However, K_c depends on test temperature, specimen thickness and constraint. A typical dependence is shown in figure 5.1. The form of this dependence has been discussed in section 3.6 and reasonably well explained in section 4.7.

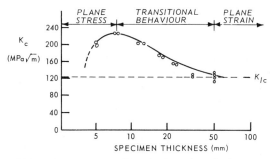

Figure 5.1. Effect of thickness on K_c behaviour for a high strength maraging steel.

Beyond a certain thickness, when the material is predominantly in plane strain and under maximum constraint, the value of K_c tends to a constant lower limit, K_{Ic}, the plane strain fracture toughness. K_{Ic} may be considered a material property, but does depend on the test temperature and loading rate. After considerable study and experimental verification the American Society for Testing and Materials (ASTM) published a standard method for K_{Ic} testing. This method will be described in section 5.2.

Of more general interest is the establishment of a test methodology for K_c determination, since the operating temperatures, loading rates and thicknesses of most materials used in actual structures are generally such that transitional plane strain-to-plane stress or fully plane stress conditions exist in service. Several engineering approaches to the problem of plane stress and transitional behaviour have been proposed. Only one, the Feddersen method, has the versatility required for structural design. This approach is the subject of section 5.3.

Considerable effort has been devoted towards R-curve testing and analysis. A "Tentative Recommended Practice" for R-curve determination was issued by the ASTM in 1976 and this was followed by a standard in 1981. The determination of R-curves is discussed in section 5.4.

...le engineering approximation (Anderson's model) to ...s of yield strength and specimen thickness on fracture ...ven. Finally, in section 5.6 the practical use of K_c, K_{Ic} is summarised.

...train Fracture Toughness (K_{Ic}) Testing

...the period in which fracture toughness testing developed (late 1950s and ...1960s) the most suitable analyses for characterizing the resistance to unstable crack growth were those of LEFM. Although it was recognized that most structural materials do not behave in a purely elastic manner on fracturing, it was hoped that provided crack tip plasticity was very limited small specimens could be used to describe the situation of unstable crack growth occurring in a large structure.

Data like those in figure 5.1 showed that a fairly constant minimum value of K_c was obtained under plane strain conditions, i.e. K_{Ic}. This value appeared to be a material property. Thus it was decided to try and establish K_{Ic} values for various structural materials, in an analogous way to the establishment of mechanical properties like yield stress and ultimate tensile strength.

Under the supervision of the ASTM E-24 Fracture Committee numerous specimen designs and test methods for K_{Ic} determination were considered. During the 1960s various parameters, e.g. notch acuity, plate thickness, fracture appearance and stress levels during fatigue precracking were investigated and resulted in the development of a standardized, plane strain K_{Ic} test method using either of two standard specimens, namely the single edge notched bend (SENB) and compact tension (CT) specimens. This method was first published in 1970 and is listed in its latest (1983) version as reference 1 of the bibliography to this chapter.

The Standard K_{Ic} Specimens

The recommended standard K_{Ic} specimens are illustrated in figures 5.2 and 5.3. Round robin test programmes have shown that these specimens enable K_{Ic} values to be reproducible to within about 15 % by different laboratories.

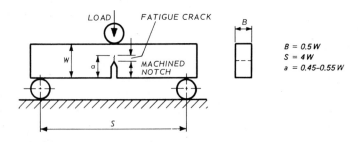

Figure 5.2. ASTM standard notched bend specimen (SENB).

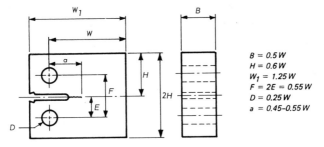

Figure 5.3. ASTM standard compact tension specimen (CT).

The stress intensity factors for these specimens are as follows: notched bend specimen (SENB)

$$K_I = \frac{LOAD \cdot S}{BW^{\frac{3}{2}}} \cdot f(\frac{a}{W}) \tag{5.1}$$

where
$$f(\frac{a}{W}) = \frac{3(\frac{a}{W})^{\frac{1}{2}}[1.99 - \frac{a}{W}(1 - \frac{a}{W})(2.15 - 3.93(\frac{a}{W}) + 2.7(\frac{a}{W})^2)]}{2(1 + 2\frac{a}{W})(1 - \frac{a}{W})^{\frac{3}{2}}}.$$

compact tension specimen (CT)

$$K_I = \frac{LOAD}{BW^{\frac{1}{2}}} \cdot f(\frac{a}{W}) \tag{5.2}$$

where
$$f(\frac{a}{W}) = \frac{(2 + \frac{a}{W})[0.886 + 4.64(\frac{a}{W}) - 13.32(\frac{a}{W})^2 + 14.72(\frac{a}{W})^3 - 5.6(\frac{a}{W})^4]}{(1 - \frac{a}{W})^{\frac{3}{2}}}.$$

The specimens must be fatigue precracked. To ensure that cracking occurs correctly the specimens contain starter notches. Several possibilities are listed in the ASTM standard, but the most frequently used is a chevron notch, figure 5.4, owing to its good reproducibility of symmetrical in-plane fatigue crack fronts.

The chevron notch forces fatigue cracking to initiate at the centre of the specimen thickness and thereby increases the probability of a symmetric crack front. After fracture toughness testing the length of the fatigue precrack at positions a_1, a_2, a_3 and at the side surfaces is measured. Assuming that other test requirements have been met the test is considered a valid K_{Ic} result if a_1, a_2 or a_3 are within 5 % of a, and if the surface crack lengths a_s are within 10% of a, where a is the mean value of a_1, a_2 and a_3.

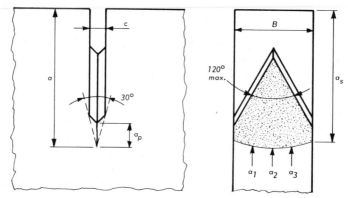

Figure 5.4. Chevron notch crack starter.

Specimen Size Requirement

The accuracy with which K_{Ic} describes the fracture behaviour depends on how well the stress intensity factor characterizes the conditions of stress and strain immediately ahead of the tip of the fatigue precrack, since it is here that unstable crack extension would originate. In establishing the specimen size requirements for K_{Ic} tests the specimen dimensions must be large enough compared with the plastic zone size, r_y, so that any effect of the plastic zone on the stress intensity analysis can be neglected and a predominantly plane strain state is ensured. The relevant dimensions (see figures 5.2 and 5.3) are:

1) The crack length, a.
2) Specimen thickness, B.
3) The remaining uncracked ligament length, W − a.

After considerable experimental work the following minimum specimen size requirements to ensure nominal plane strain behaviour were established:

$$a \geqslant 2.5 \left(\frac{K_{Ic}}{\sigma_{ys}}\right)^2$$

$$B \geqslant 2.5 \left(\frac{K_{Ic}}{\sigma_{ys}}\right)^2 \tag{5.3}$$

$$W \geqslant 5.0 \left(\frac{K_{Ic}}{\sigma_{ys}}\right)^2$$

where σ_{ys} is the yield strength. Now in chapter 3 (section 3.5) it was stated that there are empirical rules for estimating whether a specimen will be mainly in plane strain or plane stress: predominantly plane strain behaviour may be expected when the calculated size of the plane stress plastic zone, i.e. the diameter $2r_y$ in Irwin's analysis, is no larger than one-tenth of the specimen thickness. According to Irwin plane stress $\qquad r_y = \frac{1}{2\pi} \left(\frac{K_I}{\sigma_{ys}}\right)^2$ $\qquad\qquad\qquad$ (5.4)

For $K_I = K_{Ic}$ substitution of equation (5.4) in equation (5.3) shows that the minimum thickness, B, is only about 8 times the plane stress plastic zone size, $2r_y$. The empirical rule is therefore slightly conservative (a factor 10 instead of 8).

It is important to note that the specification of a, B and W (and all the other specimen dimensions) requires that the K_{Ic} value to be obtained must already be known or at least estimated. There are three general ways of sizing test specimens before the required K_{Ic} is actually obtained:

1) Overestimate K_{Ic} on the basis of experience with similar materials and empirical correlation with other types of notch toughness test, for example the Charpy impact test.
2) Use specimens that have as large a thickness as possible.
3) For high strength materials the ratio of (σ_{ys}/E) can be used according to the following table, which was drawn up by the ASTM.

σ_{ys}/E	minimum values of a and B (mm)
0.0050 − 0.0057	75.0
0.0057 − 0.0062	63.0
0.0062 − 0.0065	50.0
0.0065 − 0.0068	44.0
0.0068 − 0.0071	38.0
0.0071 − 0.0075	32.0
0.0075 − 0.0080	25.0
0.0080 − 0.0085	20.0
0.0085 − 0.0100	12.5
⩾ 0.0100	6.5

K_{Ic} Test Procedure

The steps involved in setting up and conducting a K_{Ic} test are:
1) Determine the critical dimensions of the specimen (a, B, W).
2) Select a specimen type (notch bend or compact tension) and prepare shop drawings, e.g. specifying a chevron notch crack starter, figure 5.4.
3) Specimen manufacture.
4) Fatigue precracking.
5) Obtain test fixtures and clip gauge for crack opening displacement measurement.
6) Testing.
7) Analysis of load-displacement records.
8) Calculation of conditional K_{Ic} (K_Q).
9) Final check for K_{Ic} validity.

Steps 1–3 will not be discussed further. Steps 4–6 will now be concisely reviewed, since full details may be found in reference 1 of the bibliography. The last three steps 7–9 are considered under the next subheading in this section.

The purpose of notching and fatigue precracking the test specimen is to simulate an ideal plane crack with essentially zero tip radius to agree with the assumptions made in stress intensity analyses. There are several requirements pertaining to fatigue loading. The most important is that the maximum stress intensity during the final stage of fatigue cycling shall not exceed 60 % of the subsequently determined K_Q if this is to qualify as a valid K_{Ic} result.

Recommended test fixtures for notch bend and compact tension specimen testing are described in the ASTM standard. These fixtures were developed to minimize friction and have been used successfully by many laboratories. Other fixtures may be used provided good alignment is maintained and frictional errors are minimised.

An essential part of a K_{Ic} test is accurate measurement of the crack mouth displacement (CMD) as a function of applied load. The displacement is measured with a so-called clip gauge which is seated on integral or attachable knife edges on the specimen, as shown in figure 5.5. Electrical resistance strain gauges are bonded to the clip gauge arms and are connected up to form a Wheatstone Bridge circuit, as indicated.

In carrying out the test there are requirements for specimen alignment in the test fixtures, the loading rate, test records, and post-test measurements on the specimen. The loading rate should be such that the rate of increase of stress intensity is within the range $0.55 - 2.75$ MPa$\sqrt{m}$/s. This is arbitrarily defined as 'static' loading.

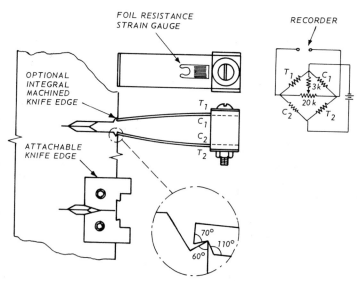

Figure 5.5. Clip gauge and its attachment to the specimen.

Each test record consists of a plot of the output of a load-sensing trans-
ducer versus the clip gauge output. It is conventional to plot load along the
y-axis, displacement along the x-axis. The initial slope of such a plot must be
between 0.7 and 1.5. The record is continued until the specimen is no longer
able to sustain a further increase in load, and the maximum load must be
determinable with an accuracy of $\pm 1\%$. Post-test measurements of the specimen dimensions, B, W, S and fatigue precrack lengths a_1, a_2, a_3, a_s (see figures
5.2–5.4) must be made to calculate K_Q and check its qualification as a valid
K_{Ic}.

Analysis of Load-Displacement Records and Determination of K_{Ic}

Plots of load versus displacement may have different shapes. The principal
types of diagram obtained are shown schematically in figure 5.6. Initially the
displacement increases linearly with load, P. In many cases there is either a
gradually increasing non-linearity, figure 5.6.a, or sudden crack extension and
arrest (called 'pop-in') followed by non-linearity, figure 5.6.b. Non-linearity
is caused by plastic deformation and stable crack growth before fast fracture.
If a material behaves almost perfectly elastically (as is rarely the case) a diagram like that in figure 5.6.c is obtained.

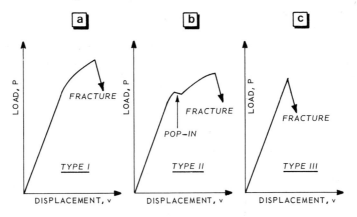

Figure 5.6. Principal types of load-displacement plots obtained during K_{Ic} testing.

To establish whether a valid K_{Ic} can be obtained from the test it is first
necessary to calculate a candidate value, K_Q, which involves a graphical construction of the load-displacement record. Then it must be determined whether
this K_Q is consistent with the specimen size and material yield strength according to equations (5.3), i.e.

$$a \geqslant 2.5 \left(\frac{K_Q}{\sigma_{ys}}\right)^2$$

$$B \geqslant 2.5 \left(\frac{K_Q}{\sigma_{ys}}\right)^2$$

$$W \geqslant 5.0 \left(\frac{K_Q}{\sigma_{ys}}\right)^2.$$

If K_Q meets these requirements then $K_Q = K_{Ic}$. If not, the test is invalid and the result may be used only to estimate the fracture toughness: it is not an ASTM standard value.

As shown in figure 5.6, the general types of P-v curves exhibit different degrees of non-linearity. Various criteria to establish the load corresponding to K_{Ic} have been considered. After considerable experimentation a 5 % secant offset was chosen to define K_{Ic} as the critical stress intensity factor at which the crack reaches an effective length 2 % greater than at the beginning of the test. Although this definition is apparently arbitrary, it turns out that for the standard SENB and CT specimens the effects of plasticity and stable crack growth are more or less accounted for by assuming an effective length increase of 2 %. (Note, however, that some more detailed remarks about the effect of stable crack growth on K_{Ic} were made in section 4.8.)

The procedure for determining the load corresponding to K_{Ic} then consists of drawing a secant line from the origin 0, with a slope 5 % less than that of the tangent 0A to the initial part of the test record. The load P_S is the load at the intersection of the secant with the test record, figure 5.7.

P_Q is defined according to the following procedure. If the load at every point on the P-displacement record which precedes P_S is lower than P_S, then P_Q is P_S (Type I, figure 5.7). However, if there is a maximum load preceding P_S that is larger than P_S, then this maximum load is P_Q (Types II and III, figure 5.7). In order to prevent acceptance of a test record for a specimen in which excessive stable crack growth occurred, it is required at this stage of the analysis that P_{max}/P_Q be less than 1.10. The value of this ratio is based on experience.

After determining P_Q the value of K_Q is calculated using appropriate stress intensity factor expressions, equations (5.1) and (5.2) for the SENB or CT specimens. The quantity $2.5\,(K_Q/\sigma_{ys})^2$ must then be found to be less than the thickness, B, and crack length, a, of the specimen for K_Q to be a valid K_{Ic}. Finally, a check is made whether the crack front symmetry requirements mentioned in the discussion to figure 5.4 are met.

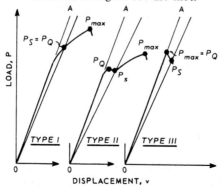

Figure 5.7. Types of load-displacement curves illustrating determination of P_S and P_Q.

5.3. Plane Stress Fracture Toughness (K_c) Testing: the Feddersen Approach

There is no standard method of plane stress fracture toughness (K_c) testing. In what follows, the engineering approach of Feddersen, which is the only method suited to practical use besides R-curve testing, will be described. The original description is given in reference 2 of the bibliography to this chapter.

Consider a thin plate under plane stress with a central crack $2a_0$ loaded in tension, figure 5.8. On reaching a stress σ_i the crack will begin to extend by slow stable crack growth. In order to maintain crack growth the stress has to be increased further: the crack will stop growing if the load is kept constant.

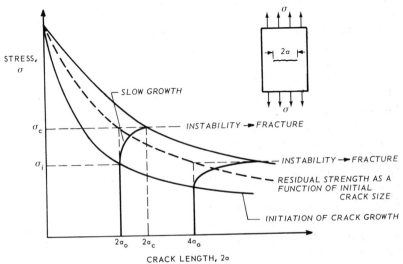

Figure 5.8. Residual strength behaviour in plane stress.

Slow crack growth continues until a critical crack size $2a_c$ is reached at a stress σ_c. Then the crack becomes unstable and fracture occurs. If the initial crack is longer, for example $4a_0$, stable crack growth starts at a lower stress, the amount of slow crack growth is larger, but σ_c is lower.

If it is assumed that all events in crack propagation and fracture are governed by the stress intensity factor, particular stress intensity factors can be attributed to each event as follows :

$$K_i = \sigma_i \sqrt{\pi a_0} \ f(\frac{a}{W})$$

$$K_c = \sigma_c \sqrt{\pi a_c} \ f(\frac{a}{W}) \tag{5.7}$$

$$K_e = \sigma_c \sqrt{\pi a_0} \ f(\frac{a}{W})$$

where K_i is the critical stress intensity for the onset of crack growth and K_c is the critical stress intensity for fracture. K_e is an apparent stress intensity, since it relates the initial crack size to the fracture stress: it has an engineering significance because irrespective of whether slow stable crack growth subsequently occurs the value of K_e defines the residual strength of a plate containing a crack of a given initial size.

(Instead of using actual crack sizes, a, in equations (5.7) the effective crack sizes $a + r_y$ could be used, as in Irwin's analysis in section 3.2. However, this is not necessary in the Feddersen approach, since in calculating stress intensity factors to obtain the residual strength diagram and the reverse operation of using the diagram to calculate the fracture stress or initial crack size the contribution of r_y cancels out.)

Tests have shown that K_i, K_c and K_e are not material constants with general validity like K_{Ic}. However, they are approximately constant for a given thickness and a limited range of crack length to specimen width, a/W. For a given material with an apparent toughness, K_e, the relation between the residual strength and initial crack size of a centre cracked panel is given ideally by the curve $\sigma_c = K_e/\sqrt{\pi a_0}$, as shown in figure 5.9.

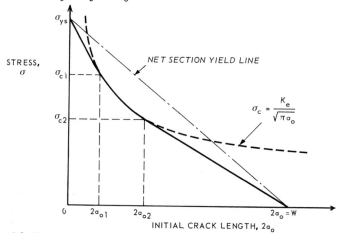

Figure 5.9. The residual strength diagram.

However, for small crack sizes σ_c tends to infinity, but the residual strength at $a_0 = 0$ cannot be larger than the yield stress. On the other hand, as $2a_0$ tends towards W, the residual strength approaches zero. At these two extremes net section yield occurs. Feddersen proposed the construction of two tangents to the curve, one from the stress axis at σ_{ys} and the other from the crack length axis at W. In the region between the points of tangency K_e is approximately constant: this piece of the curve plus the two tangents constitute the residual strength diagram.

Now a tangent to the K_e curve at any point is

$$\frac{d\sigma}{d(2a_0)} = \frac{d}{d(2a_0)}\left(\frac{K_e}{\sqrt{\pi a_0}}\right) = -\frac{\sigma}{4a_0}. \tag{5.5}$$

For the tangent from $(\sigma_{ys}, 0)$ equation (5.5) gives, see figure 5.9,

$$-\frac{\sigma_{c1}}{4a_{01}} = -\frac{\sigma_{ys} - \sigma_{c1}}{2a_{01}}, \text{ or } \sigma_{c1} = \frac{2\sigma_{ys}}{3} \tag{5.6}$$

and so the left-hand tangency point is always at $2\sigma_{ys}/3$.

Also from figure 5.9, the tangent from $(0,W)$ is defined by

$$-\frac{\sigma_{c2}}{4a_{02}} = -\frac{\sigma_{c2}}{W - 2a_{02}} \quad , \text{ or } \quad 2a_{02} = \frac{W}{3}. \qquad (5.7)$$

Thus the right-hand tangency point is always at $W/3$.

The same construction can be made for K_i and K_c, figure 5.10.

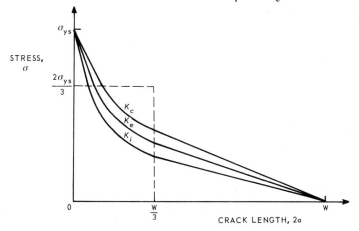

Figure 5.10. The Feddersen approach for K_i, K_e and K_c.

It has been found that the Feddersen approach represents experimental data quite well. This being so, the approach is useful because the complete residual strength diagram can be constructed for any panel size if K_e, K_c and σ_{ys} are known. Furthermore, one can easily determine the minimum panel size for valid plane stress fracture toughness testing. Consider the construction for K_e as a function of panel width, figure 5.11:

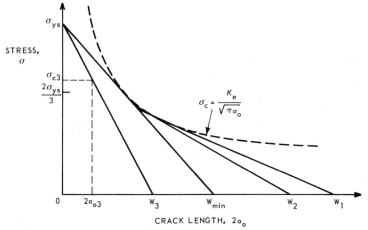

Figure 5.11. Residual strength for various panel widths.

The minimum panel size for valid K_e determination is where the two tangency points coincide. W_1 and W_2 are sufficient, but W_3 is too narrow because the failure stress lies along the line joining $(\sigma_{ys},0)$ to $(0,W_3)$ and is given by

$$\sigma_{c3} = \sigma_{ys} \left(\frac{W_3 - 2a_{03}}{W_3} \right) \tag{5.8}$$

i.e. failure always occurs at net section yield.

From $K_e = \sigma\sqrt{\pi a_0}$ the left-hand point of tangency is given by

$$K_e = \frac{2\sigma_{ys}}{3}\sqrt{\pi a_0} \ , \ \text{or} \ \ 2a_0 = \frac{9}{2\pi}\left(\frac{K_e}{\sigma_{ys}}\right)^2. \tag{5.9}$$

The two tangency points coincide when $W_{min}/3 = 2a_0$. Thus, from equation (5.9)

$$W_{min} = \frac{27}{2\pi}\left(\frac{K_e}{\sigma_{ys}}\right)^2. \tag{5.10}$$

Similarly, the value of W_{min} can be found for K_c. The screening requirements for valid plane stress fracture toughness testing follow from the foregoing considerations, namely that

$$\sigma_c < \frac{2\sigma_{ys}}{3}$$

and

$$2a < \frac{W}{3}.$$

Note that equation (5.10) cannot be used as a screening criterion. This is because violation of one of the above criteria would lead to a lower apparent K value (on a tangent rather than the residual strength curve) and substitution of this K value in equation (5.10) would give too small a value of W_{min}, i.e. an invalid test could be declared valid.

Some Experimental Considerations for K_c Testing

Actual test conditions for K_c determination are not rigorously defined as in standard tests. However, there are a number of guidelines:

1) It is obvious from what has previously been discussed that K_c tests should be conducted for material thicknesses representative of actual applications.
2) The tests should preferably be done with centre cracked panels, for which the stress intensity factors are well defined and the problems of

secondary bending and buckling are minimised (but see point 5 below).

3) Fatigue precracking is advisable, but not necessary if the notch is sharp enough to start slow crack growth well before fracture.

4) The maximum load in the test should be taken as the fracture load. Slow crack growth should be measured, preferably with a movie camera, and a load-time record synchronized with the crack growth record must be made. Also, a load-COD record is advisable.

5) In remark 2 appended to section 2.2 it was stated that a non-singular term σ_∞ must be subtracted from equation (2.10) in order to obtain σ_x for a uniaxially loaded plate, i.e.

$$\sigma_x = \frac{K_I}{\sqrt{2\pi r}} \cos \frac{\theta}{2} (1 - \sin \frac{\theta}{2} \sin \frac{3\theta}{2}) - \sigma_\infty \qquad (5.11)$$

Usually σ_∞ can be neglected. However, for $-a < x < a$ ($\theta = 180°$) the first term on the right-hand side of equation (5.11) is zero, and σ_∞ is no longer negligible. There are thus compressive stresses $\sigma_x = -\sigma_\infty$ acting along the crack surfaces. Especially in thin sheets these compressive stresses can cause local buckling near the crack, and for long cracks buckling occurs well before the specimen is ready to fail and may significantly affect the residual strength. If such buckling would be restrained in service (by structural reinforcements) then anti-buckling guides must be used in the specimen test. On the other hand, anti-buckling guides should not be used to prevent buckling that could occur in service.

5.4. The Determination of R-Curves

A significant limitation of K_c testing is that the effect of slow stable crack growth is not properly characterized. This characterization is, however, obtainable from R-curves, as has been discussed in sections 4.6 – 4.8.

An R-curve is a plot of crack growth resistance as a function of actual or effective crack extension, Δa. The crack growth resistance may be expressed in the same units as G or, as is now customary, in terms of the stress intensity factor $K_R = \sqrt{E'G}$.

R-curves can be determined by either of two experimental techniques: load control or displacement control. The load control method involves rising load tests with crack driving force (K_G) curves like those shown in figure 5.12. Under rising load conditions ($P_1 < P_2 < P_3 < P_4$) the crack extends gradually to a maximum of Δa, when unstable crack growth occurs at K_c, which is determined by the tangency point between the K_R-curve and one of the lines representing a crack driving force curve, $K_G = f(P, \sqrt{a}, a/W)$, in this case the K_G-curve corresponding to load P_4. Clearly, this testing method is capable only of obtaining that portion of the R-curve up to $K_R = K_c$, when instability occurs.

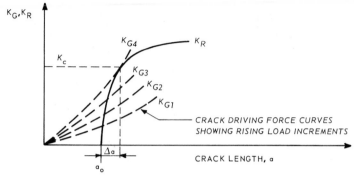

Figure 5.12. Crack growth resistance and crack driving force curves for a load controlled test.

Under displacement control a suitable specimen results in negatively sloped crack driving force curves, as shown in figure 5.13. Note that it is not sufficient to have a fixed grip condition as in figure 4.8: it is also necessary to have a specimen geometry for which dK_I/da and hence $\partial G/\partial a$ and $\partial K_G/\partial a$ are negative, i.e. K_I decreases with increasing crack length, a. Such specimen configurations are possible, for example the crack line loading case mentioned in section 2.5 and discussed later on in the present section.

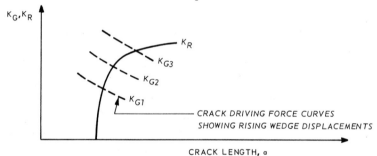

Figure 5.13. Crack growth resistance and crack driving force curves for a displacement controlled test.

Under displacement control the specimen is loaded by a wedge, which must be progressively further inserted in order to obtain greater displacements ($v_1 < v_2 < v_3$) and further crack growth. For each displacement the crack arrests when the crack driving force curve intersects the R-curve. Because there can be no tangency to the developing crack growth resistance, K_R, the crack tends to remain stable up to a plateau level, i.e. the entire R-curve can be obtained.

Relation Between R-Curve and K_c Testing

In section 4.6 it was indicated that R-curves are invariant, i.e. independent of initial crack length, a_0. However, K_c is approximately constant for only a limited range of crack lengths. The relation between R-curve and K_c testing is summarised schematically in figure 5.14. The shape of the K_c-a_0 curve in

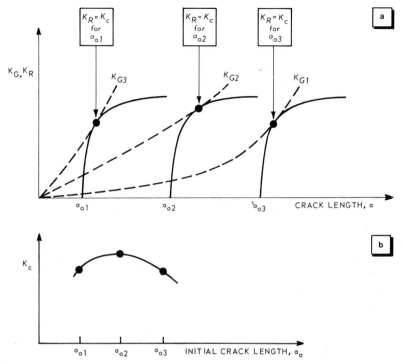

Figure 5.14. Use of R-curves for determining K_c as a function of initial crack length.

figure 5.14.b is due to two effects which partly oppose each other. First, moving the R-curve along the a axis tends to raise the (G,R) and hence (K_G, K_R) tangency points. Second, the K_G line becomes markedly curved for longer initial crack lengths, figure 5.14.a, owing to the influence of finite specimen width on the stress intensity factor, as discussed in section 2.4. Increasing curvature of the K_G lines tends to lower the (K_G, K_R) tangency points.

For given test and material conditions a K_c value (obtained by rising load testing) represents only a single point on an R-curve. But an R-curve describes the complete variation of K_c with changes in initial crack length, since R-curves are invariant. Thus an R-curve is equivalent to a large number of direct K_c tests conducted with various initial crack lengths. In practice, however, this is not too important in view of the success of the Feddersen approach, i.e. the tangency constructions to obtain the complete residual strength diagram. It is only when the estimation of the amount of stable crack growth is important that R-curve testing must be done.

Recommended Specimens for R-Curve Testing

In 1976 the ASTM published a recommended practice for R-curve determination, followed in 1981 by a standard: reference 3 of the bibliography. The ASTM method will be concisely discussed in what follows. In general the specimens will have a thickness representative of plates considered for

actual service. The ASTM recommends three types of specimens:
1) The centre cracked tension specimen (CCT).
2) The compact specimen (CS). These are the same as compact tension specimens (CT) used for K_{Ic} testing except that they may be of any thickness.
3) The crack line wedge-loaded specimens (CLWL).

The first two types of specimen are tested under load control (rising K_G curves), while the CLWL specimen may be used for displacement control tests. The specimens are illustrated in figures 5.15 – 5.17. In figure 5.17 V_1 and V_2 refer to locations at which displacements are measured in order to determine compliance and hence effective crack length. The same locations can be used for the CS specimen.

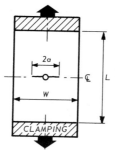

Figure 5.15. Centre cracked tension specimen (CCT).

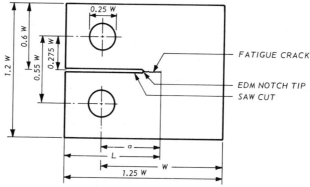

Figure 5.16. Compact specimen (CS).

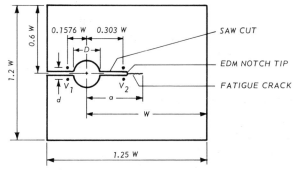

Figure 5.17. Crack line wedge-loaded specimen (CLWL).

The specimens must be fatigue precracked unless it can be shown that the machined notch root radius effectively simulates the sharpness of a fatigue precrack. For the CCT specimen the machined notch must be $30-35\%$ of W with fatigue cracks not less than 1.3 mm in length. For the CS and CLWL specimens the starter notch configuration is basically similar to that required for K_{Ic} testing, but owing to the lesser thickness a chevron notch crack starter (figure 5.4) may not be necessary to obtain a symmetrical crack front, i.e. a straight through electric discharge machined (EDM) slot will often suffice. The initial crack length must be between $35-45\%$ of W.

Specimen Size

Specimen size is based solely on the requirement that the uncracked ligaments $(W - 2a)$ or $(W - a)$ must be predominantly elastic at all values of applied load. More precisely, for the CCT specimen the net section stress based on the effective crack size $2(a_0 + \Delta a + r_y)$ must be less than the yield stress: r_y is the radius of the plastic zone according to Irwin's analysis in section 3.2. (Addition of r_y to the physical crack size is necessary because under plane stress conditions the plastic zone size is relatively large and has a significant effect on the specimen stiffness.) For the CS and CLWL specimens the condition that the uncracked ligaments must be predominantly elastic is given by the more or less empirical relation

$$W - (a_0 + \Delta a + r_y) \geqslant \frac{4}{\pi} (\frac{K_{max}}{\sigma_{ys}})^2 \tag{5.12}$$

where K_{max} is the maximum stress intensity level in the test. Equation (5.12) amounts to the requirement that the remaining uncracked ligament be at least equal to $8r_{y\,max}$.

It is worth noting here that incorporation of a plastic zone size correction will result in K_c values consistently slightly higher than those obtained by the Feddersen approach (although there is no fundamental objection to using a plastic zone size correction for the latter, see section 5.3).

R-*Curve Test Procedure*

Broadly speaking, the procedure in R-curve testing is similar to steps $1 - 7$ for K_{Ic} testing, section 5.2. However, for R-curve testing the initial step is choice of testing technique (load control or displacement control) and specimen type. The advantage of the displacement control technique in enabling determination of the entire R-curve is somewhat offset by the necessity for new or unique loading facilities, whereas with the load control method a conventional tensile machine may be used.

Additional experimental requirements are that the effective crack length must be determined and buckling prevented. The physical crack length can be measured using e.g. optical microscopy or the electrical potential method, and subsequently the measured crack length can be adjusted by the addition of r_y. Alternatively, the effective crack length can be determined directly by means of compliance measurements: the procedure for this is fully described in the ASTM standard practice. Use of compliance instrumentation also makes it possible to determine whether the specimen develops undesirable buckling despite the presence of anti-buckling guides.

All types of specimen must be loaded incrementally, allowing time between steps for the crack to stabilize before measuring load and crack length, except if autographic instrumentation is used. In the latter case the load versus crack extension can be monitored continuously, but the loading rate must be slow enough not to introduce strain rate effects into the R-curve.

To develop an R-curve the load versus crack extension data $(a_0 + \Delta a + r_y)$ can be used to calculate the crack driving force K_G, and hence K_R, using one of the two following expressions for the stress intensity factor:

CCT specimen

$$K_I = \frac{LOAD}{BW} \sqrt{\pi a \sec(\frac{\pi a}{W})}. \tag{5.13}$$

or

$$K_I = \frac{LOAD}{BW} [1.77 - 0.177(\frac{2a}{W}) + 1.77(\frac{2a}{W})^2] \tag{5.14}$$

ĊS and CLWL specimens

$$K_I = \frac{LOAD}{BW^{\frac{1}{2}}} \left[\frac{(2 + \frac{a}{W})[0.886 + 4.64(\frac{a}{W}) - 13.32(\frac{a}{W})^2 + 14.72(\frac{a}{W})^3 - 5.6(\frac{a}{W})^4]}{(1 - \frac{a}{W})^{\frac{3}{2}}} \right] \tag{5.2}$$

where B is the material thickness. Note further that for the CLWL specimen the load is indirectly obtained from a load−displacement calibration curve. The procedure for obtaining such curves is given in the ASTM standard practice, reference 3 of the bibliography.

5.5. An Engineering Approximation to Account for the Effects of Yield Strength and Specimen Thickness on Fracture Toughness: Anderson's Model

Figure 5.1. shows that the value of K_c depends on thickness, decreasing gradually to a limiting lower value of K_{Ic}. The effect of sheet thickness is related to the gradual transition from plane stress to plane strain, and this transition is strongly influenced by the yield strength, as shown schematically in figure 5.18. A higher yield strength signifies a smaller plastic zone, so that there is more material in plane strain and the fracture toughness in the transition region is lower. It is also found that K_c and K_{Ic} generally decrease with increasing yield strength.

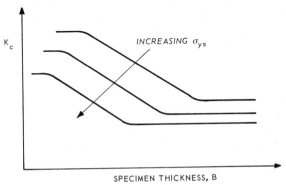

Figure 5.18. Schematic of the effects of yield strength and specimen thickness on K_c.

Although the qualitative trend shown in figure 5.18 is well established, there is no generally accepted quantitative model of the thickness effect. The simplest and also the most readily usable model is that of Anderson, figure 5.19.

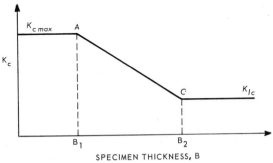

Figure 5.19. The thickness effect according to Anderson (reference 4 of the bibliography to this chapter).

The model is empirical. With knowledge of the two 'basic' fracture toughness values K_{cmax} and K_{Ic} a line is drawn between the points A and C, which can be obtained from the following empirical relations:

1) Point A is given by

$$B_1 = \frac{1}{3\pi}(\frac{K_{cmax}}{\sigma_{ys}})^2. \tag{5.15}$$

2) Point C is obtained from the limit of the ASTM condition for nominal plane strain behaviour (see section 5.2), i.e.

$$B_2 = 2.5(\frac{K_{Ic}}{\sigma_{ys}})^2. \tag{5.16}$$

The simplicity of Anderson's model makes it useful in an engineering sense, i.e. estimating fracture toughness for a practical sheet thickness when a full range of data are not available.

5.6. Practical Use of K_{Ic}, K_c and R-Curve Data

K_{Ic} data can be useful in two general ways. First, they may be used directly for choosing between materials for a particular application, especially high strength aerospace materials. More generally, since it is desirable (if possible) to avoid plane strain/low energy fracture, K_{Ic} values may be used as a basis for a screening criterion to ensure plane stress/high energy fracture. Several criteria have been proposed. One of the simplest is the through-thickness yielding criterion

$$K_{Ic} \geqslant \sigma_{ys}\sqrt{B} \tag{5.17}$$

which gives the desired increase in toughness with increasing yield strength and sheet thickness in order to obtain plane stress fracture. A full derivation of this criterion is given in reference 5 of the bibliography.

Use of the through-thickness yield criterion can be demonstrated with the help of figure 5.20, which shows $K_{Ic}-\sigma_{ys}$ data for various materials, together with lines corresponding to different ratios of K_{Ic}/σ_{ys}. From equation (5.17) it is seen that these ratio lines have the dimension $\sqrt{\text{LENGTH}}$. Thus each line may be considered to correspond to a particular sheet thickness, B, as indicated in the table in figure 5.20.

The lines give minimum values of K_{Ic}/σ_{ys} necessary for through-thickness yielding to occur in a sheet of given thickness, e.g. for a sheet 3 mm thick the minimum value of K_{Ic}/σ_{ys} is 0.055. Comparison of actual $K_{Ic}-\sigma_{ys}$ data with the ratio lines in figure 5.20 shows the following:

1) The titanium alloy beta C has high strength but low toughness. However, the alloy can still meet the through-thickness yielding criterion, equation (5.17), for sheet thickness up to 3 mm.

2) 7475 aluminium alloy and Ti-6 Al-4 V titanium alloy meet the through-thickness yielding criterion for sheet thickness up to at least 12.7 mm.

3) 10 Ni steel combines high toughness with high strength and meets the

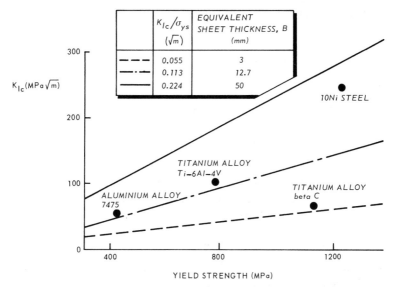

Figure 5.20. Illustration of the use of the through-thickness yielding criterion.

through-thickness yielding criterion for heavy sections approaching 50 mm thick.

Clearly, if it is required to meet the through-thickness yielding criterion in practice, then 10 Ni steel would be selected for heavy sections. On the other hand it would be possible to use beta C titanium alloy in thin sheet applications, thereby taking advantage of the lower density of titanium as compared to steel.

Intermediate plane stress – plane strain and fully plane stress fracture toughness data are primarily of interest for determining the residual strengths of actual structures using materials of the same thickness. Here again the materials may be compared, as in the introductory example given in chapter 1, section 1.8. For the majority of situations where LEFM can be applied there is little incentive to use R-curves instead of the relatively straightforward engineering approach of Feddersen. Only when the characterization of slow stable crack growth is important will R-curve data be required.

At this point in the course we come to the end of Part II, which has been concerned with LEFM. It is appropriate to note that the inability to account properly for plasticity is often a major limitation. Many engineering materials combine high toughness with low yield strength, so that the required thickness for a valid K_{Ic} test may reach the order of magnitude of a metre ! Obviously, K_{Ic} tests on such materials are neither practical nor useful, if only because the materials would never be used in such thicknesses. Also, excessive plasticity in these materials will rule out K_c testing of plates with thicknesses representative for actual structures. Resort has then to be made to Elastic-Plastic Fracture Mechanics (EPFM) characterization of the crack resistance or to a plastic collapse analysis. The subject of EPFM is treated in chapters

6—8, which comprise Part III of the course. Plastic collapse will only be mentioned in chapter 6, since it is not a fracture-dominant failure mode (see also section 1.2).

5.7. Bibliography

1. ASTM Standard E 399—83, *Standard Test Method for Plane-Strain Fracture Toughness of Metallic Materials*, 1984 Annual Book of ASTM Standards, Vol. 03.01, pp. 519—554 (1984): Philadelphia.
2. Feddersen, C.E., *Evaluation and Prediction of the Residual Strength of Center Cracked Tension Panels*, Damage Tolerance in Aircraft Structures, ASTM STP 486, pp. 50—78 (1971): Philadelphia.
3. ASTM Standard Practice E 561—81, *Standard Practice for R-Curve Determination*, 1982 Annual Book of ASTM Standards, Part 10, pp. 680—699 (1982): Philadelphia.
4. Anderson, W.E., *Some Designer-Oriented Views on Brittle Fracture*, Battelle Memorial Institute Pacific Northwest Laboratory Report SA—2290, February 11 (1969): Richland, Washington.
5. Rolfe, S.T. and Barsom, J.M., Fracture and Fatigue Control in Structures, Prentice-Hall, Inc. (1977): Englewood Cliffs, New Jersey.

Part III Elastic-Plastic Fracture Mechanics

6. SOME ASPECTS OF ELASTIC-PLASTIC FRACTURE MECHANICS

6.1. Introduction

Linear Elastic Fracture Mechanics (LEFM) was originally developed to describe crack growth and fracture under essentially elastic conditions, as the name implies. However, such conditions are met only for plane strain fracture of high strength metallic materials and for fracture of intrinsically brittle materials like glasses, ceramics, rocks and ice.

Later it was shown that LEFM concepts could be slightly altered in order to cope with limited plasticity in the crack tip region. In this category falls the treatment of fracture problems in plane stress, e.g. the R-curve concept discussed in chapter 4. Nevertheless, there are many important classes of materials that are too ductile to permit description of their behaviour by LEFM: the crack tip plastic zone is simply too large. For these cases other methods must be found.

In this part of the course we shall discuss methods in the category of Elastic-Plastic Fracture Mechanics (EPFM). These methods significantly extend the description of fracture behaviour beyond the elastic regime, but they too are limited. Thus EPFM cannot treat the occurrence of general yield leading to so-called plastic collapse. Figure 6.1 gives a schematic indication of the ranges of applicability of LEFM and EPFM in the various regimes of fracture behaviour.

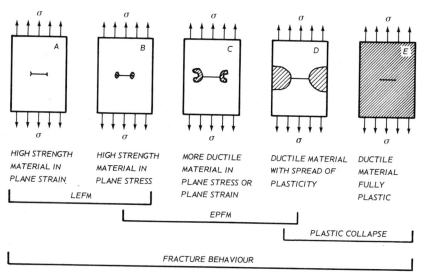

Figure 6.1. Ranges of applicability of LEFM and EPFM for describing fracture behaviour.

Since this course concerns fracture mechanics concepts, no further attention will be paid to plastic collapse, which is a yielding-dominant failure mode (see section 1.2). Discussion is confined to cases A, B, C and sometimes D of figure 6.1. The LEFM concepts applicable to cases A and B have been treated in the previous chapters 2–5. Here and in chapters 7 and 8 the principles of EPFM, which are applicable to cases B, C and D, will be given.

Note that the ranges of applicability of LEFM and EPFM overlap in figure 6.1. Before proceeding to the development of EPFM it is worthwhile to discuss these ranges of applicability in some more detail. This will be done with the help of figure 6.2.

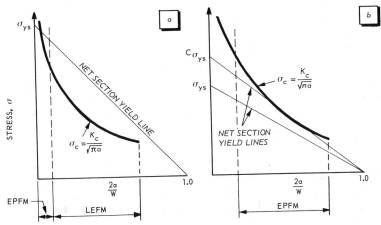

Figure 6.2. Schematic residual strength diagrams for (a) relatively brittle and (b) relatively ductile materials.

Figure 6.2.a gives a schematic residual strength diagram for a relatively brittle material in terms of the dimensionless crack length, $2a/W$ (W = panel width), of a centre cracked panel. Except for very short cracks the residual strength is determined by the stress intensity factor, since the K_c curve lies well below the line representing net section yield (and hence plasticity induced failure) of the uncracked ligaments. Thus LEFM is applicable for most cases. However, for very short cracks the plastic zone size is no longer relatively small, and EPFM concepts will have to be used.

Figure 6.2.b gives a residual strength diagram for a relatively ductile material. Clearly the unconstrained yield stress, σ_{ys}, will be reached in the uncracked ligaments well before the critical stress intensity, K_c, which is high in ductile materials. The fracture behaviour is therefore likely to be controlled by general yielding, i.e. neither LEFM or EPFM are applicable. However, in situations of high constraint, e.g. cracks in thick sections, the effective yield stress will increase to $C\sigma_{ys}$, where C is the plastic constraint factor discussed in section 3.5. The K_c curve may then predict a failure stress, σ_c, of the same order of magnitude as that given by the net section yield line: this is shown for a fairly wide range of $2a/W$ in figure 6.2.b. In such situations EPFM can be used to predict fracture behaviour. LEFM cannot be applied because σ_c will be too large a fraction of the effective yield stress and the plastic zone size will be too large.

It might be thought that situations of high constraint are rather special cases. In fact they are of prime importance with respect to practical applications. In the power generating and chemical processing industries most cracks occur in high pressure parts, which are of course thick-walled vessels and pipes. Also, the offshore industry has to cope with cracks in very large thick-sectioned welded structures. Seen in this light it is therefore not surprising that most contributions to the development of EPFM have come from these industries. In contrast LEFM is principally applied in the aerospace industry, where weight savings are at a premium and high strength, relatively brittle materials must be used.

6.2. Development of Elastic-Plastic Fracture Mechanics

Within the context of EPFM two general ways of trying to solve fracture problems can be identified:

(1) A search for characterizing parameters (cf. K, G, R in LEFM).
(2) Attempts to describe the elastic-plastic deformation field in detail, in order to find a criterion for local failure.

It is now generally accepted that a proper description of elastic-plastic fracture behaviour, which usually involves stable crack growth, is not possible by means of a straightforward, single parameter concept. On the other hand, detailed studies of elastic-plastic crack tip stress fields are most unlikely to give results suitable for practical use in the near future.

So far the only notable success of EPFM for practical applications is the ability to predict crack initiation using one or two parameters. Of the concepts developed for this purpose two have found a fairly general acceptance: the J integral and the Crack Opening Displacement (COD) approaches. Besides these concepts several others exist, e.g. the Extended Dugdale Approach and the Equivalent Energy Method, but none have received widespread recognition.

Since this course is intended to provide a basic knowledge of fracture mechanics the discussion of EPFM concepts will be limited to the generally accepted J integral and COD approaches, which are treated in sections 6.3 and 6.4 and sections 6.5–6.7 respectively. Readers interested in the less widely known concepts should consult reference 1 of the bibliography to this chapter.

6.3. The J Integral

The J integral concept is based on an energy balance approach. It was first introduced by Rice, reference 2 of the bibliography.

Consider the energy balance given in section 4.1, equation (4.1)

$$U = U_0 + U_a + U_\gamma - F. \qquad (4.1)$$

In section 4 we have only considered linear elastic behaviour. However equation (4.1) remains valid as long as the behaviour remains elastic. It need not be linear. Thus even for a material behaviour as given in figure 6.3 the equation is valid.

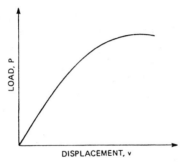

Figure 6.3. Nonlinear elastic material behaviour of a loaded uncracked plate.

An important consequence of this extended validity of equation (4.1) is that under certain restrictions this nonlinear elastic behaviour can be used to model plastic behaviour of a material. This is known as the deformation theory of plasticity. The main restriction is that no unloading may occur in any part of a body since for actual plastic behaviour the plastic part of the deformation is irreversible.

As long as equation (4.1) remains valid we can derive an instability condition as in section 4.2, leading to

$$\frac{d}{da}(F - U_a) \geqslant \frac{dU_\gamma}{da}. \tag{4.3}$$

And we can define a nonlinear elastic equivalent of G, J:

$$J = \frac{d}{da}(F - U_a). \tag{6.1}$$

Note that for elastic behaviour $J = G$ by definition. This will be discussed further in section 8.2.

Returning again to equation (4.1) we may define the potential energy U_p as

$$U_p = U_0 + U_a - F$$

i.e. $$U = U_p + U_\gamma. \tag{6.2}$$

Thus U_p contains all the energy terms that may contribute to nonlinear elastic behaviour, while U_γ (the change in elastic surface energy owing to crack extension) is generally irreversible. Since U_0 is a constant, differentiation of U_p gives

$$\frac{dU_p}{da} = \frac{d}{da}(U_a - F) = -\frac{d}{da}(F - U_a). \tag{6.3}$$

From equation (6.1) it is seen that by definition

$$J = -\frac{dU_p}{da}. \tag{6.4}$$

Now since dF/da represents the energy provided by the external force F per increment of crack extension and dU_a/da is the increase of elastic energy owing to the external work dF/da, the quantity dU_p/da represents the change in stored energy. A decrease in stored energy, $-dU_p/da$, means a release of crack driving energy, J, in order to provide the energy, dU_γ/da, for an increase in crack surface by da.

Note that during crack growth the newly formed crack flanks are completely unloaded from stresses as high as σ_{ys} (or even higher, $C\sigma_{ys}$, in the case of plane strain). Therefore in principle it should be expected that J is applicable only up to the beginning of crack growth. In fact, some success in using J to characterize crack growth has been claimed (see section 8.3 and reference 3 of the bibliography) but the topic is still somewhat controversial.

Treatment of the J integral concept will be done in four stages:

- Definition of concepts necessary to derive expressions for dU_p/da and J.
- Derivation of an expression for dU_p/da from a nonlinear energy balance.
- Definition of the J integral and proof of its path independence.
- Usefulness of the J integral concept.

It must be noted that all derivations will be made for a biaxial stress state. This simplifies the analysis but has consequences for its generality, as will be mentioned in section 6.4.

Concepts Necessary for Deriving dU_p/da and J

To understand the derivation of J the reader should be familiar with three concepts: the strain energy density, W, the traction vector, $\underline{T}$, and Green's theorem.

The strain energy density W, which is the strain energy per unit volume, is readily calculated. Consider the stress tensor σ_{ij} and the strain tensor ϵ_{ij}. For the biaxial case the indexes i and j can take the values 1 or 2, corresponding to x and y respectively; and for the variables to be considered in the derivation the following notation applies:

VARIABLE	MATHEMATICAL NOTATION	ENGINEERING NOTATION
Directions	x_1, x_2	x , y
Displacement vector	u_1, u_2	u , v
Traction vector	T_1, T_2	T_x , T_y
Stress	$\sigma_{11}, \sigma_{22}, \sigma_{12}$	$\sigma_x , \sigma_y , \tau_{xy}$
Strain	$\epsilon_{11}, \epsilon_{22}, \epsilon_{12}$	$\epsilon_x , \epsilon_y , \epsilon_{xy}$

Thus

$$\sigma_{ij} = \begin{bmatrix} \sigma_{11} & \sigma_{12} \\ \sigma_{21} & \sigma_{22} \end{bmatrix} \text{ and } \epsilon_{ij} = \begin{bmatrix} \epsilon_{11} & \epsilon_{12} \\ \epsilon_{21} & \epsilon_{22} \end{bmatrix}$$

The infinitesimal strain energy dW is given by

$$dW = \sum_{ij} \sigma_{ij}\, d\epsilon_{ij} = \sigma_{11}\, d\epsilon_{11} + \sigma_{12}\, d\epsilon_{12} + \sigma_{21}\, d\epsilon_{21} + \sigma_{22}\, d\epsilon_{22}$$

$$= \sigma_{11}\, d\epsilon_{11} + 2\sigma_{12}\, d\epsilon_{12} + \sigma_{22}\, d\epsilon_{22} \tag{6.5}$$

and

$$W = \int_0^{\epsilon_{ij}} dW = \int_0^{\epsilon_{ij}} \sigma_{ij}\, d\epsilon_{ij}. \tag{6.6}$$

The traction vector $\underline{T}$ loading a segment of a body may be defined as: $T_i = \sigma_{ij} n_j$.

$$\begin{pmatrix} T_1 \\ T_2 \end{pmatrix} = \begin{pmatrix} \sigma_{11} & \sigma_{12} \\ \sigma_{21} & \sigma_{22} \end{pmatrix} \cdot \begin{pmatrix} n_1 \\ n_2 \end{pmatrix} \quad \Rightarrow \quad \begin{aligned} \underline{T}_1 &= \sigma_{11} n_1 + \sigma_{12} n_2 \\ \underline{T}_2 &= \sigma_{21} n_1 + \sigma_{22} n_2 \end{aligned} \tag{6.7}\;\;\tag{6.8}$$

n_1 and n_2 are the cosines of the angles θ_1 and θ_2 which the vector normal to the surface of the body makes with the positive x and y directions respectively. It is easy to derive that if n is normal to a length ds of the surface then

$$dx = -ds\cos\theta_2 \quad \text{and} \quad dy = ds\cos\theta_1. \tag{6.9}$$

Green's theorem expresses a line integral along a closed path as a double integral for the area enclosed by that path. The theorem states that

$$\int_\Gamma P\, dx + Q\, dy = \iint_A \left(\frac{\partial Q}{\partial x} - \frac{\partial P}{\partial y} \right) dx\, dy \tag{6.10}$$

which in the form we are going to use can be written in parts as

$$\iint_A \frac{\partial Q}{\partial x}\, dx\, dy = \int_\Gamma Q(x,y)\, dy \tag{6.11.a}$$

and

$$\iint_A \frac{\partial P}{\partial y}\, dx\, dy = -\int_\Gamma P(x,y)\, dx. \tag{6.11.b}$$

Derivation of an Expression for dU_p/da

Consider a cracked body of unit thickness as shown in figure 6.4. The body has a perimeter Γ and a surface A. A traction $\underline{T}$ acts on a part S_0 of the perimeter and performs external work of an amount ΔF. Thus parts of the body undergo a displacement represented as a displacement vector $\underline{u}$.

Let U_{01} be the energy contained in the plate before the traction is applied. Note that U_{01} has the same meaning as U_0 in equation (4.1), except that this time we start with a plate that already contains a crack. Thus U_{01} repre-

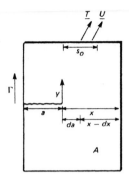

Figure 6.4. A cracked body of unit thickness loaded by a traction T.

sents the energy contained in the cracked plate owing to any previous history. The effect of applying the traction may now be considered for two cases:

- no crack growth; the potential energy is

$$U_{p1} = U_{01} + \Delta F.$$

- crack growth Δa;

$$\Delta F = \Delta U_a + \Delta U_\gamma.$$

In this case the potential energy is

$$U_{p2} = U_{01} + \Delta U_a.$$

(Note that the change in surface energy, ΔU_γ, is irreversible and cannot be part of U_{p2}.)

It follows that the change in potential energy, ΔU_p, due to a crack extension, Δa, is

$$\Delta U_p = U_{p2} - U_{p1}$$
$$= \Delta U_a - \Delta F. \qquad (6.14.a)$$

And for the limiting case $\Delta a \to 0$ we may write

$$dU_p = dU_a - dF. \qquad (6.14.b)$$

Equation (6.14.b) shows that dU_p will always be negative since dF provides both dU_a and dU_γ.

Integrating equation (6.14.b) leads to

$$\int dU_p = \int dU_a - \int dF$$

or

$$U_p = U_a - F + \text{a constant.} \qquad (6.15.a)$$

The integration constant will be equal to U_{01}, the energy content before ΔF was applied. Thus

$$U_p = U_a - F + U_{01}. \tag{6.15.b}$$

This is equivalent to the definition of U_p given at the beginning of this section. In equation (6.15.b) $U_a + U_{01}$ is the total strain energy contained in the body. This total strain energy can be represented by

$$U_a + U_{01} = \iint_A W dx dy \tag{6.16}$$

and F can be represented by

$$F = \int_\Gamma \underline{T} ds \cdot \underline{u} \tag{6.17}$$

where $\underline{u}$ is the displacement vector. Substituting equations (6.16) and (6.17) in equation (6.15.b) gives

$$U_p = \iint_A W dx dy - \int_\Gamma \underline{T} ds \cdot \underline{u}. \tag{6.18}$$

If the traction applied to the body is kept constant we may write:

$$\frac{dU_p}{da} = \iint_A \frac{\partial W}{\partial a} dx dy - \int_\Gamma \underline{T} \cdot \frac{\partial \underline{u}}{\partial a} ds. \tag{6.19}$$

Equation (6.19) is an expression for dU_p/da, the change in potential energy per unit crack extension.

Equation (6.19) can be modified as follows. As shown in figure 6.4 the coordinate system can be taken such that the origin is at the crack tip a. If the perimeter Γ is fixed, $da = -dx$ and thus $d/da = -d/dx$. Then

$$\frac{dU_p}{da} = -\iint_A \frac{\partial W}{\partial x} dx dy + \int_\Gamma \underline{T} \frac{\partial \underline{u}}{\partial x} ds. \tag{6.20}$$

Using Green's theorem, equation (6.11.a), we can eliminate A and express dU_p/da as a line integral along the perimeter Γ, i.e.

$$\frac{dU_p}{da} = -\int_\Gamma W dy + \int_\Gamma \underline{T} \frac{\partial \underline{u}}{\partial x} ds. \tag{6.21}$$

Definition of the J Integral and Proof of its Path Independence

In the beginning of section 6.3 it was shown that for nonlinear elastic behaviour we can define an energy release rate J, equation (6.1):

$$J = \frac{d}{da}(F - U_a). \tag{6.1}$$

Furthermore, it was shown that by definition, equation (6.4),

$$J = -\frac{dU_p}{da}. \tag{6.4}$$

Substituting equation (6.21) in equation (6.4) we obtain

$$J = \int_\Gamma W dy - \int_\Gamma \underline{T} \cdot \frac{\partial \underline{u}}{\partial x} ds \tag{6.22.a}$$

or

$$J = \int_\Gamma W dy - \int_\Gamma T_i \frac{\partial u_i}{\partial x} ds. \qquad (i = 1,2) \tag{6.22.b}$$

This is the definition of the J integral.

Using Green's theorem it is not difficult to prove that for a closed contour as in figure 6.5 the value of J is zero.

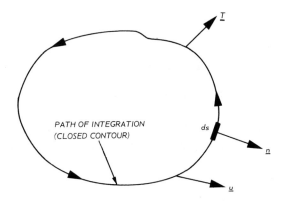

PATH OF INTEGRATION
(CLOSED CONTOUR)

Figure 6.5. A closed contour Γ with parameters used in defining the J integral.

$$J = \int_\Gamma W dy - \int_\Gamma T_i \frac{\partial u_i}{\partial x} ds \tag{6.22.b}$$

From Green's theorem, equation (6.11.a)

$$\int_\Gamma W dy = \iint_A \frac{\partial W}{\partial x} dx dy$$

$\partial W/\partial x$ is readily obtained from equation (6.5),

$$\frac{\partial W}{\partial x} = \sigma_{11} \frac{\partial \epsilon_{11}}{\partial x} + 2\sigma_{12} \frac{\partial \epsilon_{12}}{\partial x} + \sigma_{22} \frac{\partial \epsilon_{22}}{\partial x}$$

and since $\dfrac{\partial \epsilon_{11}}{\partial x} = \dfrac{\partial}{\partial x} \dfrac{\partial u_1}{\partial x}$; $2 \dfrac{\partial \epsilon_{12}}{\partial x} = \dfrac{\partial}{\partial x}(\dfrac{\partial u_1}{\partial y} + \dfrac{\partial u_2}{\partial x})$; and $\dfrac{\partial \epsilon_{22}}{\partial x} = \dfrac{\partial}{\partial x} \dfrac{\partial u_2}{\partial y}$, it follows that $\partial W/\partial x$ can be represented in terms of u_1 and u_2. Thus in general notation

$$\frac{\partial W}{\partial x} = \frac{\partial}{\partial x_j} [\sigma_{ij}(\frac{\partial u_i}{\partial x})]$$

and

$$\int_\Gamma W dy = \iint_A \frac{\partial W}{\partial x} dx dy = \iint_A \frac{\partial}{\partial x_j} [\sigma_{ij}(\frac{\partial u_i}{\partial x})] dx dy \tag{6.23}$$

Using equations (6.7) and (6.8) the second part of the right-hand side of equations (6.22.b) can be written as

$$\int_\Gamma T_i \frac{\partial u_i}{\partial x} ds = \int_\Gamma [(\sigma_{11} n_1 + \sigma_{12} n_2) \frac{\partial u_1}{\partial x} + (\sigma_{21} n_1 + \sigma_{22} n_2) \frac{\partial u_2}{\partial x}] ds$$

$$= \int_\Gamma (\sigma_{11} \frac{\partial u_1}{\partial x} + \sigma_{21} \frac{\partial u_2}{\partial x}) n_1 ds + \int_\Gamma (\sigma_{12} \frac{\partial u_1}{\partial x} + \sigma_{22} \frac{\partial u_2}{\partial x}) n_2 ds$$

From equation (6.9) $n_1\, ds = \cos\theta_1\, ds = dy$; and $n_2\, ds = -\cos\theta_2\, ds = -dx$. Thus

$$\int_\Gamma T_i \frac{\partial u_i}{\partial x}\, ds = \int_\Gamma (\sigma_{11}\frac{\partial u_1}{\partial x} + \sigma_{21}\frac{\partial u_2}{\partial x})\, dy - (\sigma_{12}\frac{\partial u_1}{\partial x} + \sigma_{22}\frac{\partial u_2}{\partial x})\, dx$$

Again using Green's theorem, this expression gives

$$\int_\Gamma T_i \frac{\partial u_i}{\partial x}\, ds = \iint_A \frac{\partial}{\partial x}(\sigma_{11}\frac{\partial u_1}{\partial x} + \sigma_{21}\frac{\partial u_2}{\partial x})\, dx\, dy + \iint_A \frac{\partial}{\partial y}(\sigma_{12}\frac{\partial u_1}{\partial x} + \sigma_{22}\frac{\partial u_2}{\partial x})\, dx\, dy$$

or

$$\int_\Gamma T_i \frac{\partial u_i}{\partial x}\, ds = \iint_A \frac{\partial}{\partial x_j}[\sigma_{ij}(\frac{\partial u_i}{\partial x})]\, dx\, dy \qquad (6.24)$$

Equation (6.24) is equal to equation (6.23). Hence

$$J = \int_\Gamma W\, dy - \int_\Gamma T_i \frac{\partial u_i}{\partial x}\, ds = 0 \qquad (6.25)$$

Thus J taken along any closed contour Γ is equal to zero.

We now take J along a closed contour in a cracked body, figure 6.6. The closed contour to be considered is ABCDEF, which includes two parts of CD and FA of the crack flanks and two paths Γ_1 and Γ_2 in opposite directions to each other.

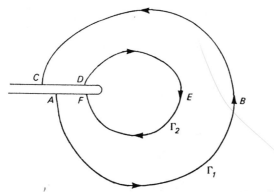

Figure 6.6. A closed contour ABCDEF for a cracked body.

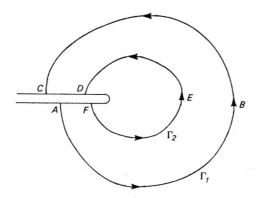

Figure 6.7. Reversing the direction of the open contour Γ_2.

Then from the knowledge that the total value of J is zero along the closed contour ABCDEF it follows that

$$J = J_{\Gamma_1} + J_{CD} + J_{\Gamma_2} + J_{FA} = 0$$

Along the crack flanks $dy = 0$ and the traction is zero, i.e. T_i is zero. Thus for CD and FA the open contours J_{CD} and J_{FA} are equal to zero. Then

$$J = J_{\Gamma_1} + J_{\Gamma_2} = 0 \; ; \; \text{or} \; J_{\Gamma_1} = -J_{\Gamma_2}$$

Reversing the direction along Γ_2 results in a change of sign. This situation is shown in figure 6.7. It follows that:

$$J_{\Gamma_1} = J_{\Gamma_2} = J \tag{6.26}$$

i.e. J is path independent when applied around a crack tip from one crack surface to another.

Usefulness of the J Integral Concept

In this section a path independent integral expression representing a non-linear elastic energy release rate has been derived. It has been stated that under certain restrictions J can be used as an elastic-plastic energy release rate.

The path independency of the J integral expression allows calculation along a contour remote from the crack tip. Such a contour can be chosen to contain only elastic loads and displacements. Thus an elastic-plastic energy release rate can be obtained from an elastic calculation along a contour for which loads and displacements are known.

Because J may be considered as an elastic-plastic energy release rate it is to be expected that there is a critical value, J_c, which predicts the onset of crack extension. This is by analogy with G_c in LEFM.

For a cracked component calculated J values can be compared to the critical value, J_c, characteristic of the material. Thus a fracture mechanics analysis can be carried out: J must always remain less than J_c.

6.4. Remarks Concerning the J Integral Concept

The J integral concept is not particularly easy to understand. However, the concept is undeniably useful, and so at this point it is also useful to direct some remarks to the derivation, applications and restrictions of J:

(1) At the beginning of section 6.3 it was stated that J would be derived assuming the deformation behaviour to be nonlinear elastic and therefore reversible. But plastic deformation is not reversible, and the energy dissipated cannot be transformed into other kinds of energy. Thus with true plastic deformation the transformation of energy terms in some

vital steps of the derivation of J are not legitimate, notably the proof
that J = 0 for a closed contour and the energy balance analysis.

(2) It was also stated that the assumption of nonlinear elasticity is compatible with actual deformation behaviour only if no unloading occurs in any part of the material. But at the crack tip the material is unloaded when crack growth occurs. Therefore J is in principle applicable only up to the beginning of crack extension and not for crack growth.

(3) In the derivation of J the strains and displacements on a body are the controlling parameters. Limitations of the derivation to a biaxial case in order to simplify the analysis means that no effects of strain in the thickness direction have been taken into account. Thus for J to be valid the strain in the thickness direction must be zero, which is only the case for plane strain and the hypothetical situation of plane strain. in a plate of unit thickness. For practical purposes this restriction is not serious, since J is primarily useful for predicting the onset of crack extension in thick sections, which are generally in plain strain.

(4) By definition J = G for the linear elastic case. Thus the J integral concept is compatible with LEFM.

Note that by analogy with G the dimensions of J are [ENERGY]/[LENGTH] per unit thickness of material, i.e. Joules/m^2 or N/m.

Further note that since J = G for the linear elastic case we may write

$$J = G = \frac{K^2}{E'} \qquad\qquad (6.27)$$

where $E' = E$ for plane stress and $E' = E/(1 - \nu^2)$ for plane strain.

(5) As stated before, it is to be expected that there is a critical material parameter, J_c, which predicts the onset of crack extension. Methods for measuring J_c will be presented in chapter 7.

(6) Obtaining solutions for the J integral in actual specimens or components turns out to be difficult. It is generally necessary to use finite element techniques. However, some simple expressions have been developed for standard specimens. They will also be presented in chapter 7.

(7) The J integral concept has been developed mainly in the USA as a fracture criterion for materials used in the power generating industry, particularly nuclear installations. In this area of application high level technology and costly production techniques are generally used and no large differences are to be expected for material behaviour in laboratory specimens and actual structures. For instance: local differences in behaviour of welded joints are not normally accounted for. This contrasts with the COD concept which is more directed to the design of welded structures, as will be discussed in section 6.7.

6.5. The Crack Opening Displacement (COD) Approach

The COD approach was introduced by Wells, reference 6 of the bibliography, as long ago as 1961. The background philosophy to the approach is as follows. In the regimes of fracture-dominant failure, cases A, B, C and partly D in figure 6.1, the stresses and strains in the vicinity of a crack or defect are responsible for failure. At crack tips the stresses will always exceed the yield strength and plastic deformation will occur. Thus failure is brought about by stresses and hence plastic strains exceeding certain respective limits.

Wells argued that the stress at a crack tip always reaches the critical value (in the purely elastic case $\sigma \to \infty$). If this is so then it is the plastic strain in the crack tip region that controls fracture. A measure of the amount of crack tip plastic strain is the displacement of the crack flanks, especially at or very close to the tip. Thus it might be expected that at the onset of fracture this Crack Opening Displacement, COD or δ_t, has a characteristic critical value for a particular material and therefore could be used as a fracture criterion.

In 1966 Burdekin and Stone provided an improved basis for the COD concept. They used the Dugdale strip yield model to find an expression for COD. Their analysis has already been reviewed in section 3.3 of the course and is given in full in reference 7 of the bibliography to this chapter. The result is

$$\delta_t = \frac{8\sigma_{ys}a}{\pi E} \ln \sec\left(\frac{\pi\sigma}{2\sigma_{ys}}\right) \tag{6.28}$$

In sections 3.2 and 3.3 it was also shown that under LEFM conditions there are direct relations between δ_t and K_I. Thus, for the Dugdale analysis

$$\delta_t = \frac{K_I^2}{E\sigma_{ys}} \tag{6.29}$$

(Note that this simple relation is obtained by taking only the first term of a series expansion of the ln sec part of equation (6.28). This is only allowed for $\sigma \ll \sigma_{ys}$.)

And for the Irwin circular plastic zone analysis

$$\delta_t = \frac{4}{\pi} \frac{K_I^2}{E\sigma_{ys}} \tag{6.30}$$

These relations are important because they show that in the elastic regime the COD approach is compatible with LEFM concepts. However, the COD approach is not basically limited to the LEFM range of applicability, since occurrence of crack tip plasticity is inherent to it.

The major disadvantage of the COD approach is that equation (6.28) is valid only for an infinite plate, and it is not possible to derive similar formulae for practical geometries. This contrasts with the stress intensity factor and J integral concepts. Thus in the first instance a characteristic value of COD at the onset of fracture can be used to compare the fracture resistance of materials but not to calculate a critical crack length in a structure. In an attempt to overcome this disadvantage the COD design curve has been developed.

6.6. The COD Design Curve

Development of the COD design curve will be reviewed in this section. More details are to be found in references 7—9 of the bibliography. The basis of the original approach to a COD design curve was that critical COD values should provide measures of the maximum permissible strains in the vicinities of cracks. Thus if a general relation could be found between COD and the local strain, then COD tests on specimens would enable assessment of the maximum permissible value of this local strain for a crack of a certain size in an actual structure. In turn this maximum permissible local strain could be compared with the actual (computed) local strain over the same distance across the crack in order to determine whether the crack were critical or not.

Although this original approach is reasonable it has the disadvantage that nothing is said about how nearly critical a crack is, nor about the maximum permissible crack size (see figure 1.4). Subsequently, however, it was found that the maximum permissible local strain could be expressed directly as a maximum permissible crack size. Computation of the local strain in a cracked structure therefore became unnecessary, and critical COD values could be directly related to maximum permissible crack sizes. This is the current approach.

Nevertheless, to properly understand the COD design curve it is best to follow its historical development, which follows.

An Analytical COD Design Curve

The starting point for the original approach to obtaining a COD design curve was the expression for δ_t in an infinite centre cracked plate. In section 3.3 it was shown that when σ/σ_{ys} is much less than unity equation (6.29) is obtained, i.e.

$$\delta_t = \frac{K_I^2}{E\sigma_{ys}} = \frac{\pi\sigma^2 a}{E\sigma_{ys}}$$

The dimension of δ_t is therefore [LENGTH]. To obtain a general parameter the COD was made dimensionless by dividing δ_t by $2\pi\sigma_{ys}a/E$, which consists of known quantities. Thus

$$\Phi = \frac{\delta_t E}{2\pi\sigma_{ys}a} = \frac{\delta_t}{2\pi\epsilon_{ys}a} \tag{6.31}$$

where $\epsilon_{ys} = \sigma_{ys}/E$, the elastic strain at the yield point. (The numerical factor of 2π was added to the denominator for convenience in a later stage of analysis.)

The next step was determination of the strain over a certain distance in a cracked plate. This is straightforward, although some complicated mathematics is involved. In section 3.3 it was shown how the COD may be derived from the non-singular stress function term $\Phi_s(z)$ in the Dugdale approach, namely by calculating the crack flank displacement v. In a similar manner it is possible to derive the displacement of the equidistant points P in figure 6.8, and hence the strain ϵ_y between these points.

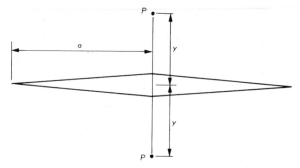

Figure 6.8. Points P distance y above and below the centre of a crack of length 2a.

Derivation of the displacement involves the same elastic displacement equation, equation (2.18.b), written in terms of μ, where $\mu = E/2(1 + \nu)$, and substituting $\nu/(1 + \nu)$ for ν, see equations (2.21.a) and (2.22.a)

$$Ev = 2\,\mathrm{Im}\,\overline{\phi}(z) - y(1 + \nu)\mathrm{Re}\phi(z).$$

However, for points not situated at the crack flank (where $y = 0$) the real part $\mathrm{Re}\,\phi(z)$ exists. The solution of this equation is beyond the scope of this course, but is given in reference 7 of the bibliography. Here only the final result for the strain ϵ_y between the points P will be given. Division by $\epsilon_{ys}(= \sigma_{ys}/E)$ gives the relative strain

$$\frac{\epsilon_y}{\epsilon_{ys}} = \frac{2}{\pi}\left[\frac{2a}{y}\,\mathrm{arc\,coth}\left(\frac{y}{a}\sqrt{\frac{(\frac{a}{a + \Delta a_n})^2 + (\frac{a}{y})^2}{1 - (\frac{a}{a + \Delta a_n})^2}}\right)\right.$$

$$\left. + (1 - \nu)\,\mathrm{arc\,cot}\sqrt{\frac{(\frac{a}{a + \Delta a_n})^2 + (\frac{a}{y})^2}{1 - (\frac{a}{a + \Delta a_n})^2}} + \nu\,\mathrm{arc\,cos}\left(\frac{a}{a + \Delta a_n}\right)\right]$$

$$(6.32)$$

where Δa_n is the Dugdale plastic zone size. For the rest of the discussion this complicated expression is fortunately not required, but it is quoted in full to show that ϵ_y/ϵ_{ys} depends on the ratio of crack length to measuring point distance, a/y.

The last step was to plot the dimensionless COD, Φ, versus the relative strain, ϵ_y/ϵ_{ys}, for several a/y values, figure 6.9. This is the original COD design curve, an analytical one. Note that there are two important limitations to the analysis besides the assumption of an infinite plate: use of the Dugdale approach implies (1) plane stress conditions and (2) elastic-perfectly plastic behaviour, i.e. no work hardening.

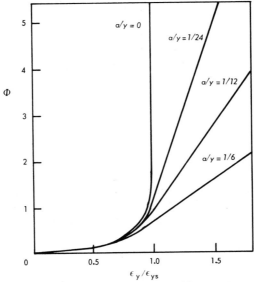

Figure 6.9. The analytical COD design curve for an infinite centre cracked plate, from reference 7 of the bibliography.

Theory and Experiment: The Empirical COD Design Curve

The intention of figure 6.9 was to provide a design curve for each a/y value such that once the critical COD was known from specimen tests the maximum permissible strain in a cracked structure could be predicted. Computation of the actual strain in the structure should then indicate whether it were in danger of failing.

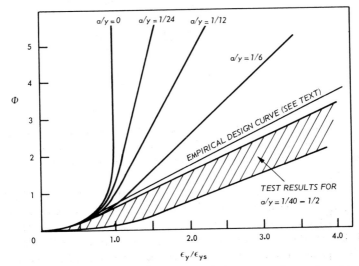

Figure 6.10. Comparison of the analytical COD design curve and experimental data, from reference 8 of the bibliography.

134

In the late 1960s many tests on wide plates were done to check the predictive capability of the COD design curve. Measurements of critical COD and maximum strain at fracture showed that the data fell into a single scatter band with no discernible dependence on a/y. Also the maximum strain at fracture was much larger than would have been predicted. These results are compared with the analytical COD design curve in figure 6.10.

Figure 6.10 shows that there is marked disagreement between theory and experiment when ϵ_y/ϵ_{ys} exceeds 0.5, which may be explained as follows. For wide plates the relative crack length, a/W, is small, such that as ϵ_y/ϵ_{ys} approaches unity the plates undergo net section or even general yielding. When net section yield occurs the assumption of elastic-perfectly plastic behaviour in the analysis becomes crucial, since material in front of the crack will in fact work harden. This results in the formation of flow bands at angles of about 45° to the crack tip, as shown schematically in figure 6.11. Once the flow bands reach the plate edges the increase in COD will be equal to the increase in *overall* displacement, and a more or less linear relation between Φ and ϵ_y/ϵ_{ys}, independent of a/y, has to be expected. Furthermore, once general yield occurs the increase in ϵ_y/ϵ_{ys} becomes much larger than the increase in Φ.

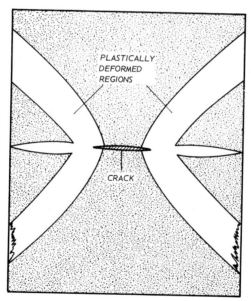

Figure 6.11. Schematic of flow bands in a wide plate.

Thus it turned out that the analytical COD design curve is useless for ϵ_y/ϵ_{ys} greater than about 0.5. The problem was obviated simply by drawing a line just above the scatter band of experimental results, thereby obtaining the empirical design curve shown in figure 6.10. This curve does in fact follow the analytical design curve for ϵ_y/ϵ_{ys} up to 0.5.

Empirical equations to describe the analytical lower part ($\epsilon_y/\epsilon_{ys} < 0.5$) and empirical upper part ($\epsilon_y/\epsilon_{ys} > 0.5$) of the COD design curve have been obtained. These are

$$\Phi = (\frac{\epsilon_y}{\epsilon_{ys}})^2 \qquad \text{for} \qquad \epsilon_y/\epsilon_{ys} < 0.5$$

$$\Phi = \frac{\epsilon_y}{\epsilon_{ys}} - 0.25 \qquad \text{for} \qquad \epsilon_y/\epsilon_{ys} > 0.5 \tag{6.33}$$

and so the COD design curve has evolved to a purely empirical correlation, even though it has an analytical background.

Note that equations (6.33) still express the dimensionless COD in terms of relative strain. From what was stated at the beginning of this section it is clearly preferable to express the COD design curve in terms of maximum permissible crack size: this is the final development leading to the current COD design curve approach.

The Empirical COD Design Curve and Maximum Permissible Crack Size

The final development of the COD design curve is due to Dawes, reference 9 of the bibliography to this chapter. Dawes argued that for small cracks ($a/W < 0.1$) and applied stresses below yield

$$\frac{\epsilon_y}{\epsilon_{ys}} \sim \frac{E\sigma}{E\sigma_{ys}} = \frac{\sigma}{\sigma_{ys}} \tag{6.34}$$

Substituting in equations (6.33) gives

$$\Phi = (\frac{\sigma}{\sigma_{ys}})^2 = \frac{\delta_t E}{2\pi \sigma_{ys} a} \qquad \text{for } \sigma/\sigma_{ys} < 0.5$$

$$\Phi = \frac{\sigma}{\sigma_{ys}} - 0.25 = \frac{\delta_t E}{2\pi \sigma_{ys} a} \qquad \text{for } \sigma/\sigma_{ys} > 0.5 \tag{6.35}$$

The maximum permissible crack size, a_{max}, can be obtained directly from equation (6.35) by substituting the critical COD value, $\delta_{t\,crit}$:

$$a_{max} = \frac{\delta_{t\,crit} E \sigma_{ys}}{2\pi \sigma_1^2} \qquad \text{for } \sigma_1/\sigma_{ys} < 0.5 \tag{6.36.a}$$

$$a_{max} = \frac{\delta_{t\,crit} E}{2\pi(\sigma_1 - 0.25\sigma_{ys})} \qquad \text{for } 0.5 < \sigma_1/\sigma_{ys} < 2 \tag{6.36.b}$$

Note the designation σ_1. This will be explained in remark (2) below.

Remarks

(1)   As mentioned earlier, use of the Dugdale approach implies plane stress conditions, i.e. the material is assumed to yield at σ_{ys}. In reality most structural parts will yield at somewhat higher stresses owing to plastic

constraint. This means that the actual COD for a crack in a structural part will be smaller than predicted, and higher stresses will be needed to reach $\delta_{t\,crit}$. Hence the COD design curve is conservative, i.e. its use will predict smaller maximum permissible strains and crack sizes than those in reality.

(2)   In equations (6.36) the designation σ_1 was introduced instead of σ. This is a design-oriented convenience: using σ_1 as the sum of all stress components (general and local) the effects of, for instance, residual stress in a weld or peak stress due to a geometrical discontinuity can be accounted for. σ_1 may reach values as high as twice σ_{ys}. More information is given in reference 8 of the bibliography to this chapter.

(3)   Equations (6.36) are also used for predicting the maximum permissible defect size of elliptical and semi elliptical defects. This is done by calculating the LEFM stress intensity factor for such defects, compare section 2.5. The result is set equal to $K = \sigma_1\sqrt{\pi a}$ for a through-thickness defect. From this equation an equivalent through thickness crack length, a, follows and this is compared to a_{max} in equations (6.36).

6.7. Further Remarks on the COD Approach

In sections 6.5 and 6.6 it has been shown that the COD approach has a sound physical and analytical background, even though the current COD design curve is empirical. Some additional remarks are given here concerning use of the COD approach.

(1)   Whether comparing the fracture resistance of materials or establishing the COD design curve it is necessary to obtain $\delta_{t\,crit}$. It has been shown experimentally that the COD depends on specimen size, geometry and plastic constraint, and so a standard COD test has been developed. This standard does not, however, precisely define the event at which δ_t is to be considered critical. There are three possibilities: δ_{ti}, the value at the onset of (stable) crack extension; δ_{tu}, the value at the point of instability; or δ_{tmax}, the value at maximum load (which is not necessarily identical to the other values, e.g. for specimens that still exhibit stable crack extension beyond the point of maximum load). The standard test is discussed further in chapter 7.

(2)   The COD approach has been developed mainly in the UK: more specifically, at the Welding Institute. The chief purpose was to find a characterizing parameter for welds and welded components of structural steels, which are difficult to simulate on a laboratory scale. Thus the COD approach is more strongly directed towards use in design of welded structures. (This of course does not mean that COD values cannot be used to compare and select materials.)

(3)   In welded steel structures the welds are most liable to fracture, not the material itself. At present the COD approach is probably the most reliable way of accounting for the fracture resistance of welds, and several weld quality specifications incorporating COD exist.

However, the issue of whether or not to use the COD approach for a number of applications is by no means settled.

(4) The British Standards Institution has published a guideline to assess the significance of weld defects (which are considered as elliptical flaws) based on the COD design curve. From service loads and measured COD values the tolerable defect sizes can be predicted. For more information the reader is referred to the official document, reference 10 of the bibliography.

(5) The COD design curve is generally considered to be conservative. For example from equation (6.36.a):

$$\delta_{t\,crit} = \frac{2\pi\sigma_1^2 a_{max}}{E\sigma_{ys}} = \frac{2K_I^2}{E\sigma_{ys}} \qquad (6.37)$$

This value of $\delta_{t\,crit}$ is twice that obtainable from the Dugdale analysis used by Burdekin and Stone, see equation (6.18), and so it should be

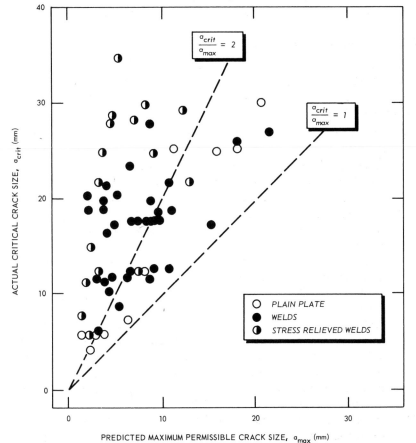

Figure 6.12. Correlation of COD design curve predictions of maximum permissible crack size with actual critical crack size for a structural steel, after reference 9 of the bibliography.

expected that at least the lower part of the COD design curve (i.e. up to $\epsilon_y/\epsilon_{ys} = \sigma_1/\sigma_{ys} = 0.5$) has a safety factor of 2. As a check on this conservatism figure 6.12 correlates COD design curve predictions of maximum permissible crack sizes, based on small specimen tests, with experimentally determined critical crack sizes in wide plates. It is seen that a safety factor of 2 bounds most of the data, all of which lie above the $a_{crit}/a_{max} = 1$ line.

6.8. Bibliography

1. Turner, C.E., *Methods for Post-Yield Fracture Safety Assessment*, Post Yield Fracture Mechanics, ed. D.G.H. Latzko, Applied Science Publishers, pp. 23–210 (1979): London.
2. Rice, J.R., *A Path Independent Integral and the Approximate Analysis of Strain Concentration by Notches and Cracks*, Journal of Applied Mechanics, Vol. 35, pp. 379–386 (1968).
3. Paris, P.C., *Fracture Mechanics in the Elastic-Plastic Regime*, Flaw Growth and Fracture, ASTM STP 631, pp. 3–27 (1977): Philadelphia.
4. Hutchinson, J.W., Nonlinear Fracture Mechanics, Technical University of Denmark Department of Solid Mechanics (1979): Copenhagen.
5. Rice, J.R., *Mathematical Analysis in the Mechanics of Fracture*, Fracture, An Advanced Treatise, ed. H. Liebowitz, Academic Press, Vol. 2, pp. 192–311 (1968): New York.
6. Wells, A.A., *Unstable Crack Propagation in Metals: Damage and Fast Fracture*, Proceedings of the Crack Propagation Symposium Cranfield, The College of Aeronautics, Vol. 1, pp. 210–230 (1962): Cranfield, England.
7. Burdekin, F.M. and Stone, D.E.W., *The Crack Opening Displacement Approach to Fracture Mechanics in Yielding*, Journal of Strain Analysis, Vol. 1, pp. 145–153 (1966).
8. Dawes, M.G., *Fracture Control in High Yield Strength Weldments*, The Welding Journal, Research Supplement, Vol. 53, pp. 369s–379s (1974).
9. Dawes, M.G., *The COD Design Curve*, Advances in Elasto-Plastic Fracture Mechanics, ed. L.H. Larsson, Applied Science Publishers, pp. 279–300 (1980): London.
10. Britisch Standards Institution PD 6493: 1980, *Guidance on Some Methods for the Derivation of Acceptance Levels for Defects in Fusion Welded Joints*, BSI (1980): London.

7. EPFM TESTING

7.1. Introduction

In chapter 6 the two most widely known concepts of Elastic-Plastic Fracture Mechanics, the J integral and Crack Opening Displacement (COD) approaches, were discussed in general terms.

This chapter will deal with test methods for obtaining values of J and COD, including critical values J_{Ic} and $\delta_{t\,crit}$. The chapter may be considered the EPFM counterpart of chapter 5, which discussed LEFM test methods.

The greater complexity of the J integral concept as compared to the COD concept is clearly demonstrated by the derivations in chapter 6. This difference in complexity is also found in the test methods. Therefore the discussion of J integral testing is subdivided into three sections:

(1) The original J_{Ic} test method, section 7.2.
(2) Alternative methods and expressions for J, section 7.3.
(3) The standard J_{Ic} test, section 7.4.

The original J_{Ic} test method requires a large amount of data analysis. This problem led to the development of certain types of test specimen for which simple expressions for J could be derived, and ultimately to the standard J_{Ic} test.

The COD concept is much more straightforward, at least from the experimental point of view. Thus only the standard COD test will be described, namely in section 7.5.

With respect to standard test methods, it has already been remarked in sections 6.4 and 6.7 that the J integral concept was developed mainly in the USA and the COD concept in the UK. Consequently it is logical that the standard for J_{Ic} is American while the COD test is the subject of an official Britisch Standard, see references 1 and 2 of the bibliography to this chapter.

7.2. The Original J_{Ic} Test Method

The first experimental method for determining J (more specifically J_{Ic}, the critical value at the onset of crack extension) was published by Begley and Landes in 1971, reference 3 of the bibliography. The method is based on the definition of J as $-dU_p/da$ per unit thickness of material, and requires graphical assessment of dU_p/da. The method will be illustrated with the help of figure 7.1, which schematically gives the graphical procedure for obtaining J_{Ic}.

The procedure is as follows:

(1) Load-displacement diagrams are obtained for a number of specimens pre-cracked to different crack lengths (a_1, a_2, a_3 in figure 7.1.a). Areas under the load-displacement curves represent the energy per unit thickness, U_1, delivered to the specimens. Thus the shaded area in figure 7.1.a is equal to the energy term U_1 for a specimen with crack length a_3 loaded to a displacement v_3.

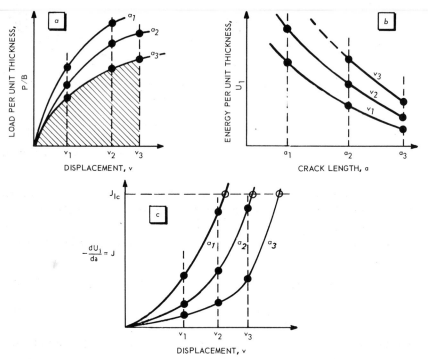

Figure 7.1. The graphical procedure involved in J_{Ic} testing according to Begley and Landes.

(2) U_1 is plotted as a function of crack length for several constant values of displacement, figure 7.1.b.

(3) The negative slopes of the U_1—a curves are plotted against displacement for any desired crack length between the shortest and longest used in testing, figure 7.1.c. The slopes represent $-(\partial U_a/\partial a)_v$ and since $(\partial U_p/\partial a)_v = (\partial U_a/\partial a - \partial F/\partial a)_v$ where $(\partial F/\partial a)_v$ equals zero they also represent $(\partial U_p/\partial a)_v = J$. Hence figure 7.1.c gives J-v calibration curves for particular crack lengths.

(4) Knowledge of the displacement v at the onset of crack extension enables J_{Ic} to be found from the J-v calibration curve for each initial crack length. In figure 7.1.c the value of J_{Ic} is schematically shown to be constant as, ideally, it should be if J is an appropriate criterion for crack extension.

The graphical procedure involves a large amount of data manipulation and replotting in order to obtain J-v calibration curves and hence J_{Ic}. There are thus many possibilities for errors, and so easier methods have been looked for, as will be discussed in section 7.3. However, the elegance of this original test method, in making direct use of the energy definition of J, remains and it is still used as a reference to check more recent developments.

7.3. Alternative Methods and Expressions for J

The main contribution to seeking alternatives for the Begley and Landes method was made by Rice et al., reference 4 of the bibliography. Their ana-

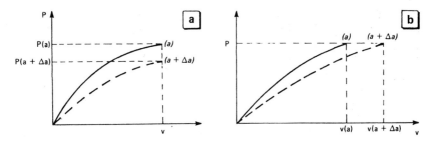

Figure 7.2. A nonlinear elastic cracked body under constant displacement (a) and under constant load (b).

lysis leads to simple expressions for J for certain types of specimen. However, before these expressions can be discussed it is necessary to consider alternative definitions of J.

Consider the expressions for U_p and J given in equations (6.2) and (6.4) respectively

$$U_p = U_0 + U_a - F \tag{6.2}$$

$$J = -\frac{dU_p}{da}. \tag{6.4}$$

For the case of constant displacement, see figure 7.2.a, we may write for U_p of a cracked body with crack length a:

$$U_p(a) = \int_0^v P(a)dv - F + U_{01}. \tag{7.1}$$

If the crack extends by an amount Δa we may write

$$U_p(a + \Delta a) = \int_0^v P(a + \Delta a)dv - F + U_{01}. \tag{7.2}$$

Note that F remains constant for the case of constant displacement.

The change in potential energy ΔU_p for a crack extension Δa is:

$$\Delta U_p = \int_0^v P(a + \Delta a)dv - \int_0^v P(a)dv = \int_0^v \Delta P \, dv \tag{7.3.a}$$

or

$$dU_p = \int_0^v dP \, dv. \tag{7.3.b}$$

Note that ΔP and dP are negative.

From equation (6.4) it follows that

$$J = -\frac{dU_p}{da} = -\int_0^v \left(\frac{\partial P}{\partial a}\right)_v dv. \tag{7.4}$$

For the case of constant load, figure 7.2.b, U_p for a body with a crack of length a can be expressed again as

$$U_p(a) = \int_0^v P(a)dv - F$$

$$= \int_0^v P(a)dv - P \cdot v = -\int_0^P v(a)dP. \tag{7.5}$$

If the crack increases by an amount Δa we may write

$$U_p(a + \Delta a) = \int_0^{v+\Delta v} P\,dv - P \cdot (v + \Delta v) = -\int_0^P v(a + \Delta a)dP. \tag{7.6}$$

The change in potential energy ΔU_p per crack extension Δa is

$$\Delta U_p = -\int_0^P v(a + \Delta a)dp - (-\int_0^P v(a)dP) = -\int_0^P \Delta v\,dP \tag{7.7.a}$$

or

$$dU_p = -\int_0^P dv\,dP. \tag{7.7.b}$$

It then follows that

$$J = -\frac{dU_p}{da} = \int_0^P \left(\frac{\partial v}{\partial a}\right)_P dP. \tag{7.8}$$

Thus the alternative definitions of J are

$$J = -\int_0^v \left(\frac{\partial P}{\partial a}\right)_v dv = \int_0^P \left(\frac{\partial v}{\partial a}\right)_P dP \tag{7.9}$$

Note the change of sign in going from constant displacement to constant load conditions. This is analogous to the change of sign for G, equation (4.23).

Using equation (7.9) Rice et al. showed that it is possible to determine J_{Ic} from a single test of certain types of specimen. As an example, J for a deeply cracked bar in bending is given straightforwardly by

$$J = \frac{2}{B(W - a)} \int_0^{\theta_c} M\,d\theta_c \tag{7.10}$$

where M is the bending moment and θ_c is the contribution of the introduced crack to the total bending angle θ_t.

Equation (7.10) is important, since it applies to a basic cracked configuration. The derivation is given in some detail here with the help of figure 7.3. M' is the bending moment per unit thickness, i.e. $M' = M/B$, and it is assumed that plastic deformation will occur only in the ligament b. Also, since $b = W - a$ it follows that $\partial/\partial b = -\partial/\partial a$.

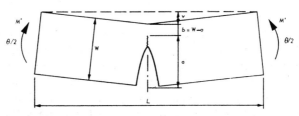

Figure 7.3. A deeply cracked bar in bending.

The total bending angle θ_t due to M' is

$$\theta_t = \theta_{nc} + \theta_c$$

where θ_{nc} represents the bending of the uncracked specimen and θ_c is the additional rotation caused by the presence of the crack. With the assumptions that all plastic deformation is confined to the ligament b and that $b \ll W$ we can express v as $\theta_t l/4$ and the load P per unit thickness as $4M'/l$. Equation (7.9) can therefore be written as

$$J = \int_0^P \left(\frac{\partial v}{\partial a}\right)_P dP = \int_0^{M'} \left(\frac{\partial \theta_t}{\partial a}\right)_{M'} \cdot dM' = -\int_0^{M'} \left(\frac{\partial \theta_t}{\partial b}\right)_{M'} \cdot dM' \tag{7.12}$$

Two more assumptions are now made. First, since the ligament b is assumed to be critical it is reasonable also to assume that $\theta_c = f(M'/b^2)$, which depends only on material properties (E, σ_{ys}, work hardening exponent n). Second, $\theta_{nc} \ll \theta_c$, so that $\theta_t \approx \theta_c$. Then

$$-\left(\frac{\partial \theta_t}{\partial b}\right)_{M'} \approx -\left(\frac{\partial \theta_c}{\partial b}\right)_{M'} = -\left(\frac{\partial f(M'/b^2)}{\partial b}\right)_{M'} = +\frac{2M'}{b^3} f'\left(\frac{M'}{b^2}\right) \tag{7.13}$$

The unknown $f'(M'/b^2)$ is determined from differentiating $f(M'/b^2)$ with respect to M'. Thus

$$\left(\frac{\partial f(M'/b^2)}{\partial M'}\right)_b = f'\left(\frac{M'}{b^2}\right) \cdot \frac{1}{b^2} \tag{7.14}$$

and so

$$f'\left(\frac{M'}{b^2}\right) = b^2 \left(\frac{\partial f(M'/b^2)}{\partial M'}\right)_b = b^2 \left(\frac{\partial \theta_c}{\partial M'}\right)_b \tag{7.15}$$

Substituting equation (7.15) in equation (7.13) gives

$$-\left(\frac{\partial \theta_c}{\partial b}\right)_{M'} = \frac{2M'}{b} \left(\frac{\partial \theta_c}{\partial M'}\right)_b \tag{7.16}$$

and

$$J = \int_0^{M'} \frac{2M'}{b} \left(\frac{\partial \theta_c}{\partial M'}\right)_b dM' = \frac{2}{B(W-a)} \int_0^{\theta_c} M \, d\theta_c \tag{7.10}$$

For a J_{Ic} test the bending moment M is plotted against θ_c. The area under the curve up to θ_{cr}, the value of θ_c at the moment of crack tension, corresponds to $B(W-a)J_{Ic}/2$. Hence in principle J_{Ic} can be determined from one test only. However, obtaining an M versus θ_c diagram is still cumbersome (though less so than the Begley and Landes method). Accordingly, Rice argued that if plasticity is restricted to the ligament $b = (W-a)$ then only a small error is made if equation (7.10) is rewritten as

$$J = \frac{2}{Bb} \int_0^{\theta_c} M \, d\theta_c = \frac{2}{Bb} \int_0^{v_c} P \, dv_c \tag{7.17}$$

i.e. it is assumed that only the ligament b rotates and the two halves of the bar remain straight.

Equation (7.17) effectively states that $J = 2U_c/Bb$, where U_c is the area under the P-v curve owing to introduction of the crack. However, in a J_{Ic} test the total displacement v_t (= $v_{nc} + v_c$) is measured, cf. equation (7.11). Consequently, the area under the P-v curve actually represents U_t rather than U_c.

The critical value U_{ccr} corresponding to J_{Ic} can be obtained as follows:

(1) Load a cracked specimen and obtain a load-displacement (P-v) diagram up to P_{cr} and v_{cr}, the critical values at which crack extension begins.

144

(2) Load a similar but unnotched specimen and obtain a P-v diagram.

(3) Determine the total energy U_{tcr} for the cracked bar from the area under the load-displacement diagram up to the point P_{cr}, v_{cr}.

(4) Similarly determine U_{nc} for the uncracked bar up to the point P_{cr}, v_{nc}.

(5) U_{ccr} is then $U_{tcr} - U_{nc}$.

Owing to the fact that $U_{nc} \ll U_c$ or U_{ccr} it is often simply assumed that U_c and U_{ccr} are equal to the total energies U_t and U_{tcr} respectively. Thus

$$J = \frac{2U_t}{Bb} \quad \text{and} \quad J_{Ic} = \frac{2U_{tcr}}{Bb} \tag{7.18}$$

and J or J_{Ic} can be calculated directly from the ligament length b (= W − a) and the load−displacement diagram for the cracked specimen.

Although, as has been stated, the method of Rice et al. makes it possible to determine J_{Ic} from only one test, this is not normally done. The reason is that detection of the beginning of crack extension is difficult. Consequently, a number of tests are made whereby each specimen is loaded to give a small but definite crack extension Δa. Then the values of J (which are, strictly speaking, invalid) are plotted versus Δa and extrapolated to $\Delta a = 0$ in order to obtain J_{Ic}. An example of this method is given in figure 7.4.

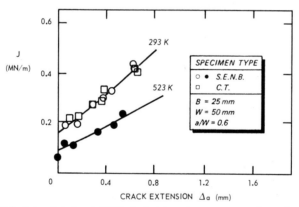

Figure 7.4. J-Δa plots for A-533B steel at two temperatures, after reference 5 of the bibliography.

J-Δa lines like those in figure 7.4 are called J resistance curves, by analogy with the LEFM R-curve. This is slightly misleading, since J is strictly valid only up to the beginning of crack extension and not beyond it. However, attempts have been made to predict stable crack extension using J resistance curves. The most promising development is the tearing modulus concept, which uses the slope dJ/da of the J resistance curve. The tearing modulus concept is discussed in section 8.3, although it must be said that its practical use is subject to many restrictions.

The J integral expression in equation (7.18) and the multiple specimen method just described form the basis for the standard J_{Ic} test, which is discussed in the next section.

7.4. The Standard J_{Ic} Test

A proposal for a standard J_{Ic} test was published in 1979. This proposal has become an ASTM Standard and was first published as such in 1982 under the designation ASTM E 813, reference 1 of the bibliography.

The J_{Ic} Specimens

The notched bend (SENB) and compact tension (CT) specimens have been selected for J_{Ic} testing, for three reasons:

(1) J is given simply by a form of equation (7.18), i.e.
    $J = (2U_t/Bb)f(a/W)$.
(2) These specimens have the same configuration as the standard K_{Ic} specimens, see section 5.2.
(3) There is considerable experience in J_{Ic} testing with these specimens.

The SENB and CT configurations have already been given in figures 5.2 and 5.3 for K_{Ic} testing. These configurations are similar for J_{Ic} testing but there are several differences in detail. Also, for both types of specimen the initial crack length a must be greater than 0.5 W to ensure that $U_{cr} \sim U_{tcr}$ (see section 7.3).

The J integral formulae for these specimens are as follows:

notched bend specimen (SENB)

$$J = \frac{2U_t}{Bb} = \frac{2U_t}{B(W-a)} \qquad (7.18)$$

compact tension specimen (CT)

$$J = \frac{2U_t}{B(W-a)} \cdot f(\frac{a}{W}) \qquad (7.19)$$

where
$$f(\frac{a}{W}) = (1 + a)/(1 + a^2)$$
and
$$a = 2\sqrt{(a/b)^2 + a/b + 1/2} - 2(a/b + 1/2).$$

J_{Ic} Test Procedure

The steps involved in setting up and conducting a J_{Ic} test are:

(1) Select a specimen type (notch bend or compact tension) and prepare shop drawings.
(2) Specimen manufacture.
(3) Fatigue precracking.
(4) Obtain test fixtures and clip gauge for crack opening displacement measurement.
(5) Testing.
(6) Analysis of load-displacement records.

(7) Determination of conditional J_{Ic} (J_Q).

(8) Final check for J_{Ic} validity.

Steps (1)-(5) will be concisely reviewed here insofar as they differ from similar steps for K_{Ic} testing in section 5.2. Steps (6)-(8) are considered under the next subheading in this section.

A special feature of J_{Ic} testing is that the clip gauge has to be positioned in the load line. For the CT specimen this means that the shape of the starter notch is different to that used in K_{Ic} testing, figure 7.5. Note that a chevron starter notch for J_{Ic} testing is not specifically recommended. This is also true for the SENB specimen. Experience has shown that a straight starter notch is usually sufficient.

Figure 7.5. CT specimen starter notches.

In order to obtain sufficiently sharp crack tips the specimen should be fatigue precracked with the maximum load not exceeding 25% of the limit load for plastic collapse P_L, which can be calculated from

SENB specimen

$$P_L = \frac{4B(W-a)^2}{3S} \cdot \frac{\sigma_{ys} + \sigma_{UTS}}{2} = \frac{4B(W-a)^2 \sigma_0}{3S} \qquad (7.20)$$

CT specimen

$$P_L = \frac{B(W-a)^2 \sigma_0}{(2W+a)} \qquad (7.21)$$

where σ_0 is called the flow stress and is typically the average of the yield strength σ_{ys} and the ultimate tensile strength σ_{UTS}. Thus $\sigma_0 = \frac{1}{2}(\sigma_{ys} + \sigma_{UTS})$. The use of σ_0 is to account for strain hardening.

The J_{Ic} tests must be carried out under controlled displacement conditions giving stable crack extension. This generally means that electrohydraulic or electromechanical testing machines are needed. As discussed in section 7.3 a number of specimens are loaded to give small but definite amounts of crack extension Δa.

The measurement of a and Δa gives specific problems. J integral test specimens are usually thick, such that "crack tunnelling" occurs during both precracking and J_{Ic} testing. This is illustrated schematically in figure 7.6. Experience has shown that to obtain consistent values of J and J_{Ic} it is necessary to take averages of at least nine measurements of a and Δa across the specimen thickness, and to count the averages of side surface crack lengths as one measurement only. To make these measurements a crack marking

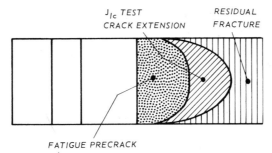

J_{Ic} TEST
CRACK EXTENSION

RESIDUAL
FRACTURE

FATIGUE PRECRACK

Figure 7.6. Schematic of a J_{Ic} test specimen broken open after testing.

technique must be employed for distinguishing between Δa and the residual fracture due to breaking open the specimen after testing. One possibility is heat tinting, i.e. heating the specimen in air to cause oxide discoloration of existing crack surfaces. Another is to fatigue cycle after the J_{Ic} test. Details of these techniques are given in reference 1 of the bibliography.

Analysis of Load-Displacement Records and Determination of J_{Ic}

The load-displacement records are analysed to obtain the areas U_c ($\sim U_t$) under the curves up to the P, v_c points corresponding to small crack extensions Δa. Values of J are then calculated by inserting U_c and averages of a into one or other of equations (7.18) and (7.19).

In the first instance these values of J are used in determining the conditional J_{Ic} (J_Q) if they satisfy the following criterion:

$$B \text{ and } W - a \text{ both} > \frac{25J}{\sigma_0}.$$

However, this criterion is not sufficient. Some J values may yet turn out to be unacceptable, depending on the amount of crack extension Δa. To check for acceptability and at the same time determine J_Q a plot similar to figure 7.4 must be constructed as shown schematically in figure 7.7.

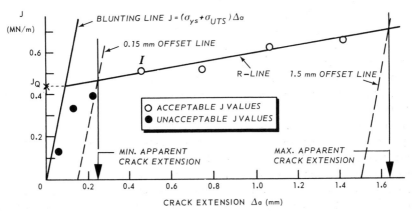

Figure 7.7. Schematic determination of acceptable J values and J_Q.

The procedure for constructing figure 7.7. is:

(1) Plot the J-Δa data points.
(2) Draw a theoretical blunting line J = $2\sigma_0 \Delta$a.
(3) Draw 0.15 and 1.5 mm offset lines parallel to the blunting line. These offset lines bound the region of acceptable J values.
(4) Draw a least squares fit line through the acceptable J-Δa points. This line is called the R-line (by analogy with the LEFM R-curve).
(5) The intersection of the R-line with the blunting line is designated J_Q. Draw two vertical lines through the intersections of the offset lines with the R-line. These vertical lines are called the minimum and maximum apparent crack extension lines.

Finally, a check is made for J_{Ic} validity. $J_Q = J_{Ic}$ if there are at least four acceptable J values, that is: between the offset lines, and if the horizontal distance of the nearest data point (I in figure 7.7) to the minimum apparent crack extension line is less than one-third the distance between the two apparent crack extension lines.

ASTM E 813 does allow an alternative determination of the R-line which enables a truly single specimen J_{Ic} determination. After loading the specimen until a small amount of crack extension occurs the load is decreased by about 10 %. In a load–displacement (P–v) diagram, figure 7.8, this is reflected as the first part of the elastic compliance line for unloading. From the elastic compliance we can calculate the instantaneous crack length a and thus Δa = a $-$ a_0, cf. section 4.4.

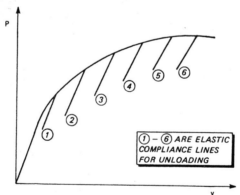

Figure 7.8. The unloading compliance technique.

ASTM E 813 contains a table equating dimensionless crack length a/W to dimensionless compliance values for the standard SENB and CT specimens.

From the values of a, Δa and P a point on the J–Δa curve can be calculated. By repeating this process a number of times an R-line can be obtained from a single specimen. A disadvantage of the method is that a accurate measurement of the unloading compliance line requires sophisticated equipment and great experimental skill. Up to now the method has found only limited application.

Background to the J_{Ic} *Determination*

(1) The minimum thickness requirement $B > 25 J/\sigma_0$ ensures that crack extension Δa occurs under plane strain. It is an empirical requirement based on tests with steels, and its purpose is to enable a straight R-line to be obtained which is readily extrapolated to J_Q. Specimens with insufficient thickness result in curved R-lines which are extremely difficult to extrapolate.

(2) The minimum ligament length requirement $b = (W - a) > 25 J/\sigma_0$ is also empirical and is intended to prevent net section yield, see section 6.1.

(3) The blunting line procedure was adopted to account for the apparent increase in crack length owing to crack tip blunting. This apparent increase in crack length will be less than or equal to the blunted crack tip radius, which in turn is half the crack opening displacement δ_t. Thus the apparent $\Delta a \leqslant 0.5\delta_t$. Assuming $\delta_t = J/\sigma_0$, a relation discussed further in point (5) and chapter 8, the apparent crack extension due to crack blunting can be accounted for by $\Delta a = 0.5\delta_t = J/2\sigma_0$, or

$$J = 2\sigma_0 \Delta a. \tag{7.22}$$

Although the concept of accounting for crack blunting is correct, the use of equation (7.22) is still criticized for two reasons, which are discussed in the next two points (4) and (5).

(4) The blunting line and the R-line are influenced by work hardening. With more work hardening the slope of the blunting line is less, while the R-line is observed to be steeper. This leads to much more potential error in estimating J_Q, as is shown in figure 7.9.

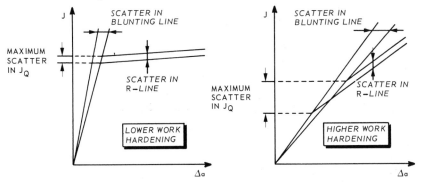

Figure 7.9. Influence of work hardening on J_Q estimation error.

(5) The blunting line, equation (7.22), is based on $J = \delta_t \sigma_0$. As will be discussed in more detail in chapter 8, relations of the form $J = M\delta_t\sigma_0$ are reasonable, but the factor M can vary between 1 and 3, and often has a value ~ 2. This means that the blunting line slope according to the ASTM standard may be too shallow, which results in an overestimation of J_Q, as figure 7.10 shows.

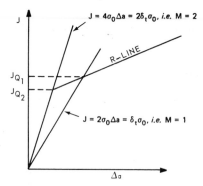

Figure 7.10. Influence of the relation between J and δ_t on J_Q.

Experiments have shown that the overestimation of J_Q may be as much as 10%, reference 7 of the bibliography.

(6) The 0.15 mm offset line ensures that Δa is at least 0.15 mm and so can be measured accurately enough. The 1.5 mm offset line ensures that Δa is generally less than 6% of the remaining ligament in the SENB and CT specimens proposed for J_{Ic} testing, and it has been shown that up to this amount of crack extension the appropriate J integral formulae, equations (7.18) and (7.19) remain valid.

(7) The final checking criteria for J_{Ic} validity have been devised to minimise scatter and improve the reliability of the J resistance curve.

Concluding Remarks

First it should be noted that the J_{Ic} test procedure according to ASTM standard E 813 only allows J_{Ic} (or J_Q) to be determined. In no way does it justify assessment of J values beyond J_{Ic} or the physical reality of J resistance curves.

Before publication of the standard J_{Ic} test some ten different procedures had been used. Chipperfield (reference 7) reviewed these methods and showed that J_{Ic} values obtained in different ways varied by up to 20%. This demonstrated the need for a standard test. Also, it is evident that J_{Ic} data from the literature should be carefully checked as to reliability before using them.

7.5. The Standard COD Test

At the beginning of this chapter it was remarked that the COD test is the subject of an official British Standard, reference 2 of the bibliography.

The Standard COD Specimen

The standard COD test specimen conforms to the notched bend (SENB) configuration already described in section 5.2. The specimen thickness B is specified to be 0.5W, although specimens with B = W may be used in special cases. In both cases B must be equal to the thickness of the material as used in service.

Expressions for Calculating δ_t

Direct measurement of δ_t at the crack tip is impossible. Instead a clip gauge is used to measure the crack opening v_g at the specimen surface, figure 7.11.a. It is assumed that the ligament b (= W − a) acts as a plastic hinge.

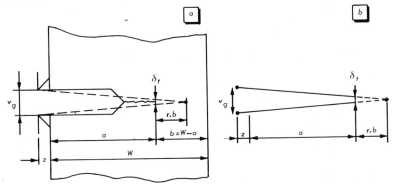

Figure 7.11. Definition of surface crack opening v_g and the relation to δ_t.

This implies a rotation point within the ligament at some distance r·b Figure 7.11.b shows that δ_t can be expressed as

$$\delta_t = \frac{v_g r \cdot b}{r \cdot b + a + z} \tag{7.23}$$

where the distance z corrects for the position of the clip gauge if attachable knife edges are used (see figure 5.5).

Although equation (7.23) is simple, there are two notable difficulties:

(1) The value of the rotation factor r. Experimentally determined values of r lie between 0.33 and 0.48, and a nominal value of 0.4 has been assigned for the standard COD test. This is because determination of r requires complicated techniques, e.g. the double clip gauge method (reference 8) or infiltration of the crack with plastic or silicone rubber (reference 9).

(2) Interpretation of the clip gauge displacement v_g. The increase in v_g with loading from a null point setting is caused by two effects, namely elastic opening of the crack and rotation around r·b. Thus to consider v_g as arising only from rotation, as in equation (7.23), would lead to erroneous results. v_g must be separated into an elastic part v_e and a plastic part v_p as shown schematically in figure 7.12.

Only the plastic part of the displacement is substituted into equation (7.23), i.e.

$$\delta_p = \frac{v_p r b}{r \cdot b + a + z}. \tag{7.24.a}$$

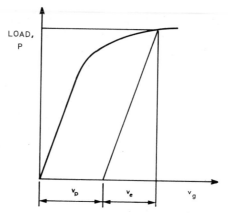

Figure 7.12. Separation of total displacement v_g into elastic (v_e) and plastic (v_p) components.

The elastic part v_e is not used but the elastic contribution to δ_t is calculated according to the LEFM expression for COD, equations (3.16) and (6.29), modified for plane strain and a plastic constraint factor C = 2 (see also section 3.5). Thus

$$\delta_e = \frac{K_I^2}{E\sigma_{ys}}\left(\frac{1-\nu^2}{2}\right) \tag{7.24.b}$$

and

$$\delta_t = \delta_e + \delta_p = \frac{K_I^2(1-\nu^2)}{2E\sigma_{ys}} + \frac{v_p rb}{rb + a + z}. \tag{7.25}$$

Note that the value of K_I in equation (7.25) is obtained from the standard formula for the SENB specimen, equation (5.1), by substituting the initial crack length a and the load at which v_p is measured.

As will be seen under the subheading "Analysis of Load-Displacement Records to Determine δ_t," several values of δ_t can be defined. Note that the British Standard defines δ_t as the crack opening at the original crack tip, as shown in figure 7.11.b. This means that it is taken for granted that during loading the crack tip will displace and move forward owing to blunting, since at the very tip δ_t must always be zero.

COD *Test Procedure*

The steps involved in setting up and conducting a COD test are:
(1) Prepare shop drawings of the SENB specimen.
(2) Specimen manufacture.
(3) Fatigue precracking.
(4) Obtain test fixtures and clip gauge for crack opening displacement measurement.
(5) Testing.
(6) Analysis of load-displacement records to determine δ_t.

Steps (1), (2) and (4) will not be considered further in view of previous discussions in section 5.2. Steps (3) and (5) will be reviewed here and step (6) will be dealt with under the next subheading.

The configuration of the starter notch for fatigue precracking is similar to that for the SENB K_{Ic} specimen, see section 5.2, except that a straight notch is recommended rather than a chevron. Fatigue precracking has to be done with a stress ratio R (= $\sigma_{min}/\sigma_{max}$) between 0 and 0.1, and K_{max} may not exceed $0.63 \sigma_{ys} \sqrt{B}$. These requirements are to ensure a sufficiently sharp precrack with limited residual plastic strain in the crack tip region. Note that equation 5.1 may be used to calculate stress intensity factors.

The requirements for precrack length a_p, total crack length a, and the measurement of total crack length are exactly the same as in K_{Ic} testing, see section 5.2 and in particular the discussion of figure 5.4. In fact, it is possible in some cases to obtain a valid K_{Ic} from a COD test. This will be returned to in the next subsection.

COD tests can be carried out with any testing machine incorporating a load cell to measure force electrically. The British Standard specifies that the loading rate should be such that the increase in stress intensity factor with time, dK_I/dt, is between 0.5 and 2.5 MPa$\sqrt{m}$/s. This is arbitrarily defined as 'static' loading, in the same way as for K_{Ic} testing. Note that since the increase rate dK_I/dt is measured in the elastic region of the load−displacement curve this procedure can lead to large differences in loading rate for ductile specimens: if the loading rate of the testing machine is kept constant the rate of displacement will strongly increase in the plastic region of the load−displacement curve; if, on the other hand, the displacement rate of the testing machine is kept constant, the loading rate will decrease in the plastic region. It has been shown that low loading rates in the plastic region of the load− displacement diagram may lead to lower COD values, see reference 10 of the bibliography.

Analysis of Load-Displacement Records to Determine δ_t

The load-displacement records can assume five different forms. These are given schematically in figure 7.13. The assessment of δ_t for each case will be briefly discussed.

Cases I and II are treated similarly. Case I is a monotonically rising load-displacement curve showing limited plasticity and no stable crack extension. Case II shows a "pop-in" owing to sudden crack extension and arrest. In both cases P_c and v_c are used to calculate δ_{tc} according to equation (7.25). Also, the 5% secant method of determining K_{Ic} (see section 5.2) can be applied, and a valid K_{Ic} may be obtainable. This is the reason why the standard COD test requires fatigue precracking and crack length measurement to be done in the same way as for K_{Ic} testing.

Cases III and IV may also be treated similarly. In both cases stable crack extension occurs and two values of δ_t can be calculated: δ_{ti} at (P_i, v_i) for the beginning of crack extension; and δ_{tu} at (P_u, v_u) for crack instability or pop-

154

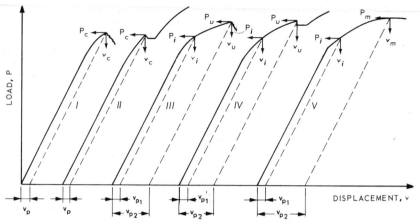

Figure 7.13. Types of load-displacement plots obtained during COD testing.

in. However, correct determination of δ_{ti} requires crack extension to be monitored for its commencement, thus giving (P_i, v_i). The British Standard (reference 2) lists various methods of crack monitoring, of which the simplest is the DC Potential Drop method.

Case V is relevant to extremely ductile materials for which stable crack extension proceeds beyond maximum load P_m. Two values of δ_t can be calculated: δ_{ti} as in cases III and IV; and δ_{tm} corresponding to (P_m, v_m).

Concluding Remarks: The Practical Significance of δ_t

Strangely enough, although the procedure for COD testing is standardised, the British Standard does not specify what value of δ_t should be used as the critical COD value $\delta_{t\,crit}$, which was discussed in sections 6.6 and 6.7. The Standard states that

"The selection of the critical value (either that at the initiation of slow crack growth or those at the onset of unstable crack extension or the first attainment of maximum force plateau) shall be by agreement between the manufacturer and purchaser".

In fact, the value of $\delta_{t\,crit}$ is usually taken to be that at instability or pop-in (δ_{tc} for cases I and II, δ_{tu} for cases III and IV) or maximum load (δ_{tm} for case V).

7.6. Bibliography

1. ASTM Standard E 813–81, *Standard Test Method for* J_{Ic}, *A Measure of Fracture Toughness*, 1982 Annual Book of ASTM Standards. Part 10, pp. 822–840 (1982): Philadelphia.
2. British Standard Institution BS 5762, *Methods for Crack Opening Displacement (COD) Testing*, BSI (1979): London.
3. Landes, J.D. and Begley, J.A., *The Influence of Specimen Geometry on* J_{Ic}, Fracture Toughness, ASTM STP 514, pp. 24–39 (1972): Philadelphia.
4. Rice, J.T., Paris, P.C. and Merkle, J.G., *Some Further Results of J Integral Analysis and Estimates*, Progress in Flaw Growth and Fracture Toughness Testing, ASTM STP 536, pp. 231–245 (1973): Philadelphia.

5. Pickles, B.W., *Fracture Toughness Measurements on Reactor Steels*, Proceedings of the 2nd European Colloquium on Fracture, Darmstadt, reported in Vortschrittsberichte der VDI Zeitschriften, Series 18, Vol. 6, pp. 130–143 (1978).

6. Blauel, J.G. and Hollstein, T., *On the Determination of Material Fracture Parameters in Yielding Fracture Mechanics*, Proceedings of the 2nd European Colloquium on Fracture, Darmstadt, reported in Vortschrittsberichte der VDI Zeitschriften, Series 18, Vol. 6, pp. 178–185 (1978).

7. Chipperfield, C.G., *A Summary and Comparison of J Estimation Procedures*, Journal of Testing and Evaluation, Vol. 6, pp. 253–259 (1978).

8. Veerman, C.C. and Muller, T., *The Location of the Apparent Rotation Axis in Notched Bend Testing*, Engineering Fracture Mechanics, Vol. 4, pp. 25–32 (1972).

9. Robinson, J.N. and Tetelman, A.S., *Measurement of K_{Ic} on Small Specimens Using Critical Crack Tip Opening Displacement*, Fracture Toughness and Slow-Stable Cracking, ASTM STP 559, pp. 139–158 (1974): Philadelphia.

10. Tsuru, S. and Garwood, S.J., *Some Aspects of Time Dependent Ductile Fracture of Line Pipe Steels*, Mechanical Behaviour of Materials, Pergamon Press, Vol. 3, pp. 519–528 (1980): New York.

8. ADDITIONAL INFORMATION ON J AND COD

8.1. Introduction

This last chapter on Elastic-Plastic Fracture Mechanics also concerns the J integral and Crack Opening Displacement (COD) concepts discussed in chapters 6 and 7. There are three topics:

- The relation between J and COD, section 8.2.
- J and stable crack growth, the tearing modulus concept, section 8.3.
- Validity and applicability of the tearing modulus concept, section 8.4.

Discussion of the relation between J and COD is of itself interesting, but it is also helpful for illustrating the equivalence of J and G in the LEFM regime and that J is compatible with LEFM principles.

Stable crack growth under EPFM conditions is not well understood. Extensive treatment of this subject is beyond the scope of this course. However, in section 8.3 the tearing modulus concept, which appears to be the most promising analysis at present, will be discussed in some detail. This leads to the last section, concerning the validity and applicability of the tearing modulus concept.

8.2. The Relation between J and COD

In chapter 6 it was remarked that the J integral and COD concepts have been developed mainly in the USA and UK respectively. In the first instance the two concepts seem to be unrelated, and until recently it was not widely known that there is a definite relation. In the late 1970s a number of expressions relating J and COD were published. They all take the form

$$J = M\delta_t\sigma_{ys} \tag{8.1}$$

where M varies between 1.15 and 2.95. General acceptance of equation (8.1) is indicated by the use of a similar expression $J = \delta_t\sigma_0$ in the blunting line procedure for J_{Ic} testing, see section 7.4.

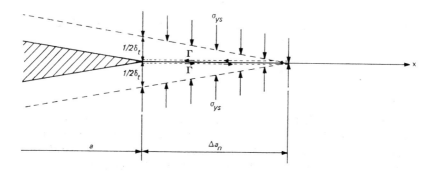

Figure 8.1. Dugdale model with closed contour Γ around the yielded strip.

Hutchinson (reference 1 of the bibliography to this chapter) showed that derivation of equation (8.1) is relatively simple when the Dugdale strip yield model is used. His analysis is therefore reproduced here, although since it uses a model it does not constitute a definite proof of the relation between J and COD.

Consider a Dugdale type crack as shown in figure 8.1 with a closed contour Γ around the yielded strip. From section 6.3 and in particular equations (6.22)–(6.25)

$$J = \int_{\Gamma} (W dy - T_i \frac{\partial u_i}{\partial x} ds) \tag{8.2}$$

or, since $dy = n_1 ds = \cos\theta_1 ds$

$$J = \int_{\Gamma} W\cos\theta_1 ds - \int_{\Gamma} \sigma_{ij} n_{ij} \frac{\partial u_i}{\partial x} ds. \tag{8.3}$$

Since J is path independent we may take the closed contour along the lower and upper sides of the yielded strip (counterclockwise in figure 8.1). Then only σ_{ys} acts on the contour. This stress is normal to the x axis and so $\cos\theta_1 = 0$.
Thus

$$J = - \int_{\Gamma} \sigma_{ij} n_{ij} \frac{\partial u_i}{\partial x} ds \tag{8.4}$$

and

$$J = - \int_{\Gamma} \sigma_{ys} n_2 \frac{\partial u_2}{\partial x} ds = \sigma_{ys} \int_{\Gamma} \frac{\partial v}{\partial x} dx. \tag{8.5}$$

(Note that $dx = -n_2 ds$, see section 6.3).

Taking J counterclockwise along Γ means proceeding along the lower side of the strip from a to $a + \Delta a_n$ in the (x^+) direction and back along the upper side from $a + \Delta a_n$ to a in the (x^-) direction. Thus

$$J = \sigma_{ys} \left[\int_a^{a+\Delta a_n} \frac{\partial v}{\partial x} dx^+ - \int_a^{a+\Delta a_n} \frac{\partial v}{\partial x} dx^- \right]$$

$$= \sigma_{ys} [v(a + \Delta a_n)^+ - v(a)^+ - v(a + \Delta a_n)^- + v(a)^-].$$

From figure 8.1 it is obvious that $v(a + \Delta a_n)^+ = v(a + \Delta a_n)^- = 0$. Also, $v(a)^+ = -\frac{1}{2}\delta_t$ and $v(a)^- = +\frac{1}{2}\delta_t$. Consequently

$$J = \sigma_{ys} [-v(a)^+ + v(a)^-] = \sigma_{ys} \delta_t. \tag{8.6}$$

Remarks

(1) Since only the definition of the strip yield model is used there is no restriction on the length of Δa_n (the plastic zone). Thus Hutchinson's analysis is valid under both LEFM and EPFM conditions. As stated earlier, the analysis is not a definite proof since it uses a model. However, it clearly demonstrates the existence of a relation between J and COD of the type given in equation (8.1).

(2) More complex analyses, often using finite element calculations, also give results of the form $J = M\delta_t\sigma_{ys}$, where M is a function of σ_{ys}/E and the strain hardening exponent n, see references 2 and 3 of the bibliography.

(3) Experimental values of M are often ~ 2, e.g. figure 8.2, in contrast to equation (8.6), where M = 1. The higher values of M found experimentally are most probably due to the real plastic zone behaving differently from that assumed by the Dugdale approach.

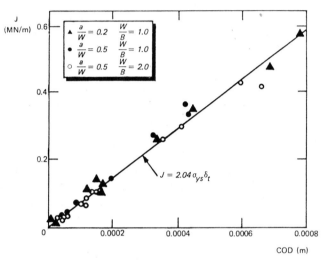

Figure 8.2. Experimental relation between J and COD for notched bend (SENB) tests of a steel with $\sigma_{ys} = 370\,\mathrm{MPa}$. After reference 4.

(4) Returning to the validity of equation (8.6) under LEFM conditions, we can use the information at the end of section 3.3, where it was shown that for σ/σ_{ys} much less than unity

$$\delta_t = \frac{K_I^2}{E\sigma_{ys}} \quad ; \text{i.e.} \quad \delta_t = \frac{G}{\sigma_{ys}}. \tag{8.7}$$

Substitution into equation (8.6) gives

$$J = \delta_t\sigma_{ys} = G. \tag{8.8}$$

Thus it can be illustrated that J is indeed equal to G in the LEFM regime.

8.3. J and Stable Crack Growth: the Tearing Modulus Concept

In sections 6.3 and 7.3 it was stated that the J integral concept is strictly valid only up to the beginning of crack growth. However, J shows a well-defined rise with increasing crack extension Δa, e.g. figure 7.4, and this has resulted in $J - \Delta a$ plots being referred to as J resistance curves, and to the definition of an R-line for J_{Ic} testing, figure 7.7.

A related aspect is that it may be highly conservative (i.e. inefficient) to use J_{Ic} as a measure of the fracture resistance to be expected in practice.

This is because the R-line for many materials has a very steep slope, and only a few millimetres of stable crack extension give J values two or three times J_{Ic}.

It is therefore no surprise that attempts are being made to describe stable crack growth under elastic-plastic conditions. So far only a few of these attempts have received notable attention. One is the tearing modulus concept developed by Paris et al., reference 5. This concept appears to be the most promising, and so it will be treated in some detail here.

P.C. Paris is generally credited with developing the tearing modulus concept. However, J.R. Rice and J.W. Hutchinson have made important contributions, as Paris acknowledges in his publications on the subject.

The Tearing Modulus Concept

Paris et al. stated that crack extension by blunting and stable tearing could be interrupted by unstable crack growth due to cleavage either before or after the initiation of ductile crack growth, depending on the material. In the present context only the more ductile behaviour, i.e. stable tearing without cleavage is of interest.

Figure 8.3, which is a schematic J resistance curve, shows the various possibilities. Note that the resistance curve for stable crack growth is represented by a straight line. Such behaviour is observed for a number of materials at least during the first few millimetres of crack extension.

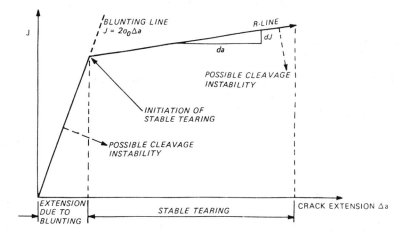

Figure 8.3. A schematic $J-\Delta a$ diagram.

The $J-\Delta a$ curve in figure 8.3 represents a material parameter for crack resistance which is a function of crack extension Δa only and is independent of the initial crack length. On the other hand, if a structural component containing a crack is loaded, a certain value of J will be applied and can be considered as a 'crack driving force'. This J value will depend mainly on the load level, crack length and the geometry of the component. For a large number of geometries such J values can be calculated, either by analytical

160

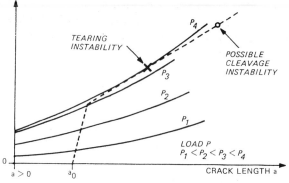

Figure 8.4. The crack driving force diagram.

or by numerical methods, such as finite element techniques. The results of these calculations can be represented in J–a diagrams, as shown schematically by the solid lines in figure 8.4.

Since the material parameter J is independent of the initial crack length, the crack initiation and stable crack growth behaviour of a component containing a crack of length a_0 can now be studied by inserting the material's J resistance curve (J–Δa) starting at this initial crack length a_0.

From figure 8.4 it may be observed that after initiation at load level P_3 the load may increase to P_4 before tearing instability occurs. Also the amount of stable crack growth can be assessed from figure 8.4.

In the recent literature on stable crack growth, for instance reference 6 of the bibliography, figure 8.4 is referred to as a 'crack driving force diagram'. Note the analogy with the LEFM R-curve approach. If the slope of the J–a curve applied by the loading system exceeds the slope of the material's J–Δa curve, instability occurs.

Stability Assessment: T_{mat} and T_{app}

A serious drawback of the approach adopted in figure 8.4 is its lack of accuracy. The J value and amount of crack extension have to be obtained from a graph.

Paris suggested to calculate the slopes of both curves and compare the values thus obtained. For the straight stable crack growth part of the resistance curve of figure 8.3 this calculation is fairly simple:

$$\frac{dJ}{da} = C_1 \qquad (8.9)$$

which, however, is temperature dependent for most materials since the J–Δa diagram depends on temperature. Paris found that dividing J by σ_0 (the flow stress) results in diagrams reasonably independent of temperature. Thus a temperature independent description of stable tearing is given by

$$\frac{dJ}{da} \cdot \frac{1}{\sigma_0} = C_2 \qquad (8.10)$$

where σ_0 is typically the average of σ_{ys} and σ_{UTS}.

Since the $J-\Delta a$ diagram and σ_0 are material characteristics, equation (8.10) has a unique value for a given material and therefore fully describes its stable tearing behaviour. Instead of equation (8.10) Paris introduced the parameter

$$T_{mat} = \frac{dJ}{da} \cdot \frac{E}{\sigma_0^2} \qquad (8.11)$$

where T stands for 'tearing modulus'. This value of T_{mat} has to be compared to a similar expression, T_{app}, which is the tearing modulus applied to a cracked body by the loading system:

$$T_{app} = \left(\frac{dJ}{da}\right)_{app} \frac{E}{\sigma_0^2} . \qquad (8.12)$$

As soon as $T_{app} \geqslant T_{mat}$ the occurrence of instability must be expected.

Calculating the slope of the $J_{app}-$crack length a curves is a very complicated task and for most geometries it is impossible to solve the problem analytically. An exception is the SENB specimen under limit load condition.

For the interested reader the analysis is given below. (The analysis was taken from reference 5 of the bibliography.)

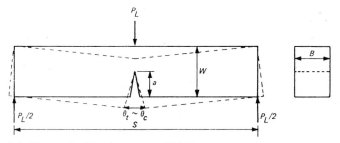

Figure 8.5. The notched bend specimen (SENB) under limit load.

We assume that all plasticity is confined to the remaining ligament b (= W − a) and that the loading system is rigid, giving a fixed displacement. The bending moment at limit load is

$$M_L = \frac{P_L}{2} \cdot \frac{S}{2} = A\sigma_0 Bb^2. \qquad (8.13)$$

For plane strain A ≈ 0.35 (see reference 6).

If the crack extends under fixed displacement the load drops. Therefore since $P = (4A\sigma_0 Bb^2)/S$,

or

$$\frac{dP_L}{da} = -\frac{dP_L}{db} = \frac{4A\sigma_0 B}{S}(-2b)$$

$$dP_L = \frac{4A\sigma_0 B}{S}(-2b\,da) \qquad (8.14)$$

If the load drops from P_L the elastic rotation of the beam θ_{el} (= $PS^2/12EI$) decreases. But since the total displacement is fixed, there has to be a compensating increased rotation θ_p of the plastic ligament of the cracked bar, i.e. $d\theta_t = (d\theta_{el} + d\theta_p) = 0$. Thus

$$d\theta_p = -d\theta_{el} \sim -\frac{4dv}{S} = -\frac{dP_L S^2}{12EI}. \qquad (8.15)$$

Note that θ_{el} is approximated by $4v/S$, which is only valid for a deeply cracked bar with plasticity restricted to the ligament, see also section 7.3, the derivation of equation (7.17). Since $I = BW^3/12$

$$d\theta_c = -\frac{dP_L S^2}{EBW^3} \, .$$ (8.16)

Substitution of equation (8.14) into equation (8.16) gives

$$d\theta_p = \frac{8A\sigma_0 Sb\, da}{EW^3} \, .$$ (8.17)

The expression for J given by equation (7.10) is

$$J = \frac{2}{Bb} \int_0^{\theta_c} M d\theta_c$$

where θ_c again represents the bending angle of the cracked bar, and so

$$dJ = \frac{2M}{Bb} d\theta_c \, .$$ (8.18.a)

This means that an increase in θ_c causes an increase in J. On the other hand, crack extension results in a decrease in J (cf. figure 7.2). This decrease has been approximated by Paris as

$$dJ = -\theta_c \sigma_0 \, da.$$ (8.18.b)

Thus in total

$$dJ = \frac{2M}{Bb} d\theta_c - \theta_c \sigma_0 \, da.$$ (8.18.c)

Finally, since an increase in θ_c corresponds to an increase of θ_p, we may substitute equations (8.13) and (8.17) in equation (8.18) to give

$$dJ = \frac{16A^2 \sigma_0^2 b^2 S\, da}{EW^3} - \theta_c \sigma_0 \, da$$ (8.19)

which can be rearranged to

$$T_{app} = \frac{dJ}{da} \cdot \frac{E}{\sigma_0^2} = \frac{16A^2 b^2 S}{W^3} - \frac{\theta_c E}{\sigma_0}$$ (8.20)

In reference 7 Paris et al. applied the tearing modulus concept to notched bend bars of a rotor steel in the following way:

(1) Obtain specimens with different a/W and load span S. This enables instability conditions to vary.
(2) Carry out tests at 303 and 503 K to ensure ductile behaviour (i.e. no premature interruption of stable crack growth by cleavage) and also provide a check on the temperature independence of T_{mat} and T_{app}. Δa is derived from an analysis of compliance, which is determined from load-displacements records. θ_c is also derived from displacement via the approximate relationship $\theta_c = 4v/S$.
(3) Calculate T_{app} according to equation (8.20) and using A = 0.353.
(4) Approximate T_{mat} by

$$T_{mat} = \frac{\Delta J}{\Delta a} \cdot \frac{E}{\sigma_0^2}$$ (8.21)

where $\Delta J = 2\Delta U/Bb$. Equation (8.21) is based on the proposed standard expression for the SENB specimen, equation (7.16).
(5) Compare T_{app} and T_{mat} as shown in figure 8.6. Note that T is non-dimensional.

Figure 8.6 includes the line $T_{app} = T_{mat}$ which, if the concept of tearing modulus is correct, should represent the boundary between stable and unstable crack growth. The experimental data provide good confirmation of this and also a reasonably consistent value of T_{mat}.

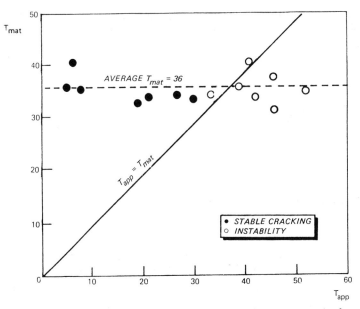

Figure 8.6. Experimental comparison of T_{mat} and T_{app} for a rotor steel, after reference 7.

If both T_{app} for a certain structural component and T_{mat} for the component material are calculated, it is customary to represent the result in a single J–T diagram as shown schematically in figure 8.7. In recent literature on the tearing modulus concept such a diagram is referred to as 'the stability assessment diagram'.

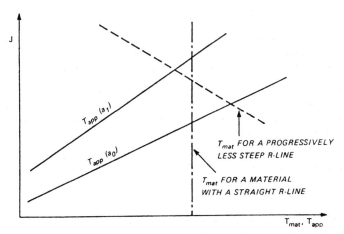

Figure 8.7. A schematic stability assessment diagram. $T_{app} \geqslant T_{mat}$ indicates tearing instability.

Since T_{mat} is a characterizing parameter for a certain material the J–T_{mat} curve is a fixed line in the stability assessment diagram. If the J resistance line for stable crack growth is straight as in figure 8.4, T_{mat} will be a constant independent of J. For a number of structural steels, however, the slope of the J-resistance line becomes less steep with increasing crack extension. This will result in a negative slope of the $J-T_{mat}$ diagram.

The applied tearing modulus will strongly depend on load, crack length and compliance of the structural component studied. Therefore it is possible to calculate various T_{app} curves, for instance for different initial crack lengths a_0 and a_1 as in figure 8.7. From knowledge of J_{Ic} at initiation and the value of J at instability, obtained by means of the stability assessment diagram, the amount of stable crack growth can be established.

8.4. Validity and Applicability of the Tearing Modulus Concept

Since the J integral concept is strictly valid only up to the beginning of crack extension it is necessary to examine carefully the validity regime of the tearing modulus. A thorough discussion is given in reference 8 of the bibliography. An outline will be given here.

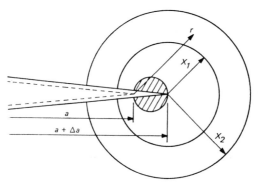

Figure 8.8. Different regions around an elastic-plastic crack tip.

We may distinguish three regions around the tip of a crack extending under elastic-plastic conditions. As illustrated in figure 8.8 these regions are:

(1) The crack extension region Δa in which elastic unloading occurs. J is invalid in this region.

(2) A region of size X_1 in which nonproportional plastic loading occurs. This cannot be described by the deformation theory of plasticity, and so J is invalid also in this region.

(3) A region of size X_2 in which nearly proportional loading occurs and is determined by the deformation theory of plasticity. In this region a J analysis would be valid.

Consideration of these three regions leads to the following requirements for valid application of J under stable crack growth conditions:

$$\Delta a \ll X_2$$

X_1 small in absolute terms

X_1 small relative to X_2.

The second and third requirements are readily fulfilled if the J contour is taken at a distance r from the initial crack tip, see figure 8.8, such that

$$\frac{da}{r} \ll \frac{dJ}{J}. \tag{8.22}$$

The first requirement is dealt with by restricting the amount of stable crack extension to the first few millimetres for which J increases rapidly with Δa to give a straight R-line. More specifically, a boundary condition is applied as shown schematically in figure 8.9. This condition limits the amount of stable crack extension to a value D for which the R-line rises to a J value twice that at the initiation of stable crack growth. Since the R-line is effectively straight, any point on it will give

$$\frac{J}{D} = \frac{dJ}{da} \tag{8.23}$$

or

$$\frac{1}{D} = \frac{dJ}{da} \cdot \frac{1}{J}. \tag{8.24}$$

Combining equations (8.22) and (8.24) gives

$$\frac{1}{r} \ll \frac{dJ}{da} \cdot \frac{1}{J} = \frac{1}{D}$$

or

$$D \ll r.$$

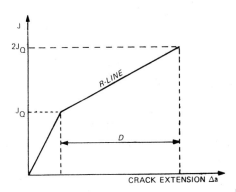

Figure 8.9. Restriction of the amount of stable crack extension.

$D \sim \Delta a$, which is much less than X_2, and r has to be less than X_2. Thus J integral analysis may be considered valid if

$$D \ll r < X_2. \tag{8.25}$$

Furthermore, X_2 will be a fraction of the uncracked ligament b (= W − a). Thus the condition in equation (8.25) may be written simply as

$$D \ll b \quad \text{or} \quad \frac{1}{D} \gg \frac{1}{b}$$

166

which leads to

$$\frac{dJ}{da} \cdot \frac{1}{J} \gg \frac{1}{b}$$

or
$$\frac{dJ}{da} \cdot \frac{b}{J} = \omega \gg 1. \qquad (8.26)$$

Equation (8.26) expresses the condition for valid application of J and hence the validity regime of the tearing modulus. This condition is, however, vague with respect to the minimum value of ω for which the J integral and tearing modulus concepts are still valid. At present a minimum value of ω cannot be specified. (Note that for the tests discussed in section 8.3 ω had a value ~ 15.)

Applicability of the Tearing Modulus

Having shown that the tearing modulus concept is valid with certain restrictions (small crack extension and straight R-line) its usefulness in practice will be considered.

First it must be noted that criteria for initiation of crack extension, e.g. K, G, J and COD, are based only on stress, crack length and a material parameter. In general, no further information is needed for predicting initiation of fracture in a component or structure. On the other hand, expressions for T_{app} are very complex.

Consider the expression for T_{app} for the SENB specimen under limit load conditions, obtained in the small print part of section 8.3:

$$T_{app} = \frac{16A^2 b^2 S}{W^2} - \frac{\theta_c E}{\sigma_0}. \qquad (8.20)$$

The expression depends on crack length (b = W − a), loading (θ_c), specimen size, or more generally, specimen compliance (S, θ_c), and the strain hardening characteristics of the material under consideration ($\sigma_0 = (\sigma_{ys} + \sigma_{UTS})/2$). This is complicated enough and yet equation (8.20) was derived via a simplified analysis under limit load conditions and is only valid for one specific specimen geometry.

Besides the foregoing, it has been shown that the J resistance curve, and in particular the R-line for stable crack growth, is strongly dependent on the state of stress (i.e. plane strain, plane stress) and material anisotropy.

Thus it is evident that the practical use of the tearing modulus concept is subject to many restrictions. Nevertheless, as stated in section 8.1, it is the most promising analysis method for stable crack growth that is currently available. The work presented in reference 9 of the bibliography is the first step towards practical application of the tearing modulus concept.

For a number of standard geometries J_{app}–crack length a solutions are presented in such a way that only standard material properties (such as E, σ_0 and strain hardening exponent n) have to be substituted to obtain the

J_{app}—a curve for a cracked component of a specific material under a given load.

If the $J-\Delta a$ curve of such a material is known, the crack driving force diagram as well as the stability assessment diagram can be constructed.

Such a catalogue of standard solutions could serve a similar purpose as the LEFM stress intensity factor handbooks which were discussed in section 2.7 (references 3 and 4 of chapter 2).

8.5. Bibliography

1. Hutchinson, J.W., Non Linear Fracture Mechanics, Technical University of Denmark Department of Solid Mechanics (1979): Copenhagen.
2. Tracey, D.M., *Finite Element Solutions for Crack Tip Behaviour in Small Scale Yielding*, Transactions ASME, Journal of Engineering Materials and Technology, 98, pp. 146—151 (1976).
3. Rice, J.R. and Sorensen, E.P., *Continuing Crack Tip Deformation and Fracture for Plane Strain Crack Growth in Elastic-Plastic Solids*, Journal of the Mechanics and Physics of Solids, 26, pp. 163—186 (1978).
4. Dawes, M.G., *The COD Design Curve*, Advances in Elasto-Plastic Fracture Mechanics, ed. L.H. Larsson, Applied Science Publishers, pp. 279—300 (1980): London.
5. Paris, P.C., Tada, H., Zahoor, A. and Ernst, H., *The Theory of Instability of the Tearing Mode of Elastic Plastic Crack Growth*, Elastic-Plastic Fracture, ASTM STP 668, American Society for Testing and Materials, pp. 5—36 (1979): Philadelphia.
6. Shih, C.F., German, M.D. and Kumar, V., International Journal of Pressure Vessel and Piping, 9, pp. 159—196 (1981).
7. Paris, P.C., Tada, H., Ernst, H. and Zahoor, A., *Initial Experimental Investigation of Tearing Instability Theory*, Elastic-Plastic Fracture, ASTM STP 668, American Society for Testing and Materials, pp. 251—265 (1979): Philadelphia.
8. Hutchinson, J.W. and Paris, P.C., *Stability Analysis of J Controlled Crack Growth*, Elastic-Plastic Fracture, ASTM STP 668, American Society for Testing and Materials, pp. 37—64 (1979): Philadelphia.
9. Kumar, V., German, M.D. and Shih, C.F., *An Engineering Approach for Elastic Plastic Fracture Analysis*, Report Prepared for Electric Power Research Institute, EPRI NP 1931 (1981): Palo Alto, California.

Part IV   Fracture Mechanics Concepts for Crack Growth

9. FATIGUE CRACK GROWTH

9.1. Introduction

Engineering components and structures often operate under alternating loads sufficiently severe to make fatigue resistance a primary design criterion. In other words, the designer must ensure that a component or structure has an adequate fatigue life. This is by no means straightforward: the fatigue life comprises both crack initiation and propagation stages, as shown in figure 9.1, and an exact definition of the transition from initiation to propagation is usually not possible. The most that can be said is that crack initiation and microcrack growth generally account for most of the life, especially in the low stress–long life regime.

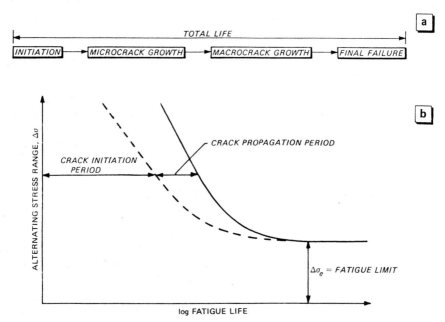

Figure 9.1. Schematic representation of the fatigue life and its dependence on stress levels.

Of basic importance in the context of this fracture mechanics course is the division into microcrack and macrocrack growth periods. This division can be defined in various ways. An apparently very reasonable definition is that a macrocrack has dimensions sufficient for its growth to depend only on bulk properties and conditions rather than on local ones. In LEFM terms this means that macrocrack growth can be described by the stress intensity factor concept. The merits of such an approach will be discussed in section 9.2.

As in the case of statically loaded cracks there are plastic zones at the tips of propagating fatigue cracks. These plastic zones have a significant effect on crack growth, to be discussed in section 9.3. Another important influence is the effect of the environment. An overview of environmental effects on fatigue crack growth is given in section 9.4. However, the effects of material microstructure are not dealt with in this chapter. Instead, they form part of chapter 13, sections 13.3 and 13.4.

Sections 9.2—9.4 are concerned with the straightforward application of fracture mechanics to fatigue crack growth. In section 9.5 this application is extended to predicting crack growth under constant amplitude loading. In sections 9.6 and 9.7 fatigue crack propagation under variable amplitude loading and methods for its prediction are discussed.

Finally, in section 9.8 the special problem of crack growth from notches will be treated briefly. At the roots of notches and other geometrical discontinuities the stress may exceed the yield stress of the material. The growth of small cracks (small compared to the notch dimensions) in such regions will be controlled by local notch plasticity and application of LEFM is either impossible, in which case EPFM must be used, or possible only with substantial modifications and restrictions.

9.2. Description of Fatigue Crack Growth Using the Stress Intensity Factor

The main question concerning fatigue crack growth is: how long does it take for a crack to grow from a certain initial size to the maximum permissible size, i.e. the crack size at which failure of a component or structure is just avoidable. This is one of the five basic questions posed in section 1.3. There are three aspects to this question:

- the initial crack size, a_d
- the maximum permissible or critical crack size, a_{cr}
- the period of crack growth between a_d and a_{cr}.

The initial crack size, a_d, corresponds to the minimum size that can be reliably detected using non-destructive inspection (NDI) techniques. The maximum permissible crack size, a_{cr}, can be determined, at least in principle, using LEFM or EPFM analysis to predict the onset of unstable crack extension. The third aspect requires knowledge of a fatigue crack growth curve, schematically shown in figure 9.2. Note that there is an initial flaw size, a_0, which is not the same as a_d and is non-zero. This is because real components and structures either contain certain discontinuities (voids, flaws, damage, inhomogeneities) or a microcrack initiates at a smooth surface and grows to a size big enough for fracture mechanics to apply, but still too small for detection.

Experimental determination of fatigue crack growth curves for every type of component, loading condition, and crack size, shape and orientation in a structure is impractical, not to say impossible. Fortunately, at least for constant amplitude loading, one can use the correlation between fatigue crack growth rate, da/dn, and the stress intensity factor range, ΔK, already discussed in section 1.9. For example, suppose the relation between da/dn and ΔK is known from standard tests. Then, provided the stress intensity — crack length relationship can be determined for a component, it is possible to specify da/dn for each crack length, and the required a—n curve can be constructed by inte-

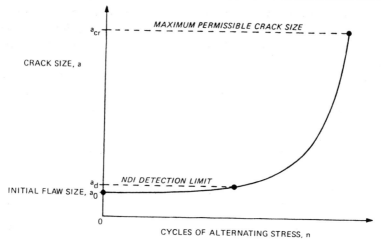

Figure 9.2. Schematic fatigue crack growth curve.

grating the standard $da/dn - \Delta K$ data over the appropriate range of crack length (a_d to a_{cr} in figure 9.2).

The Fatigue Crack Growth Rate Curve $da/dn - \Delta K$

The characteristic sigmoidal shape of a $da/dn - \Delta K$ fatigue crack growth rate curve is shown in figure 9.3, which divides the curve into three regions according to the curve shape, the mechanisms of crack extension and various influences on the curve. In region I there is a threshold value, ΔK_{th}, below which cracks do not propagate. Above this value the crack growth rate increases relatively rapidly with increasing ΔK. In region II there is often a linear log-

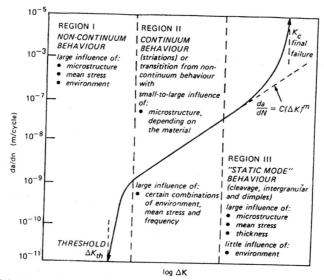

Figure 9.3. Characteristics of the fatigue crack growth rate curve $da/dn - \Delta K$.

log relation between da/dn and ΔK. Finally, in region III the crack growth
rate curve rises to an asymptote where the maximum stress intensity factor,
K_{max}, in the fatigue stress cycle becomes equal to the critical stress intensity
factor, K_c.

There have been many attempts to describe the crack growth rate curve by
"crack growth laws", which usually are semi or wholly empirical formulae
fitted to a set of data. The two most widely known are

$$\frac{da}{dn} = C(\Delta K)^m \qquad \text{— the Paris equation} \qquad (9.1)$$

$$\frac{da}{dn} = \frac{C(\Delta K)^m}{(1-R)K_c - \Delta K} \qquad \text{— the Forman equation} \qquad (9.2)$$

Paris' equation describes only the linear log-log (region II) part of the crack
growth curve, as indicated in figure 9.3. Forman's equation also describes
region III.

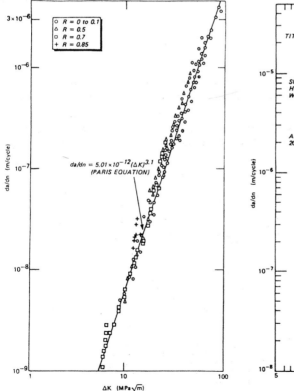

Figure 9.4. Fatigue crack growth for structural
steel BS4360 at room temperature and with
cycle frequencies 1–10 Hz. R is the stress
ratio $\sigma_{min}/\sigma_{max}$, cf. section 1.9.

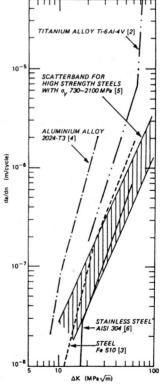

Figure 9.5. Fatigue crack growth
rates as functions of ΔK for
various structural materials at
low R values.

It is also possible to describe the complete da/dn$-\Delta$K curve by an expression like

$$\frac{da}{dn} = C(\Delta K)^m \left[\frac{1 - \left(\dfrac{\Delta K_{th}}{\Delta K}\right)^{n_1}}{1 - \left(\dfrac{K_{max}}{K_c}\right)^{n_2}}\right]^{n_3} \tag{9.3}$$

where n_1, n_2 and n_3 are empirically adjusted exponents.

The significance of such equations is limited, but they can be useful in providing a first estimate of crack growth behaviour, especially if the material concerned exhibits region II behaviour over a wide range of crack growth rates. An example is given in figure 9.4. The material is a higher strength structural steel often used for offshore structures, reference 1 of the bibliography to this chapter.

Figure 9.5 gives an impression of the crack growth rate behaviour of a number of well known structural materials for low values of the stress ratio, R = $\sigma_{min}/\sigma_{max}$: the numbers in square brackets are references in the bibliography. The positions of the crack growth rate curves for the various types of material represent a general trend. Thus aluminium alloys generally have higher crack propagation rates than titanium alloys or steels at the same ΔK values, and the data for steels fall within a surprisingly narrow scatter band despite large differences in composition, microstructure and yield strength.

The ability of ΔK to correlate crack growth rate data depends to a large extent on the fact that the alternating stresses causing crack growth are small compared to the yield strength. Therefore crack tip plastic zones are small compared to crack length even in very ductile materials like stainless steels.

However, even though they are small, fatigue crack plastic zones can significantly affect the crack growth behaviour. This will be shown in the next section.

9.3. The Effects of Stress Ratio and Crack Tip Plasticity: Crack Closure

The stress ratio R, which is the ratio of the minimum and maximum stresses in a fatigue cycle, can have a significant influence on the crack growth behaviour. As stated in section 1.9, it has been found experimentally that

$$\frac{da}{dn} = f(\Delta K, R). \tag{9.4}$$

In other words, besides the stress intensity factor range, ΔK, there is an influence of the relative values of K_{max} and K_{min}, since R = $\sigma_{min}/\sigma_{max}$ = K_{min}/K_{max}. This is illustrated in figure 9.6, which shows that crack growth rates at the same ΔK value are generally higher when R is more positive.

There is no immediately obvious explanation for the effect of R. A proper explanation requires the understanding of crack closure, to be discussed next.

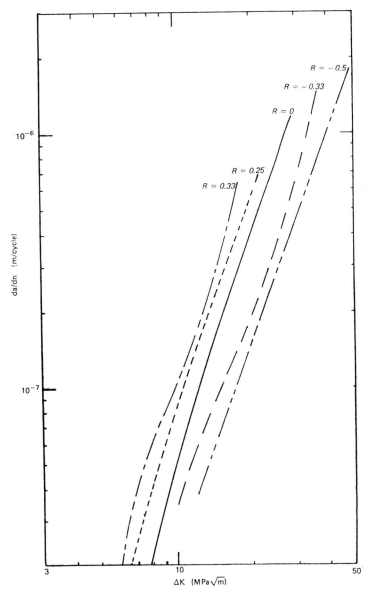

Figure 9.6. Influence of R on fatigue crack growth in aluminium alloy 2024-T3 Alclad sheet.

However, it is worth noting here that the effect of R has proved to be strongly material dependent, a fact readily observed by comparing figures 9.4 and 9.6.

Crack Tip Plasticity and Crack Closure

In the early 1970s Elber (reference 7 of the bibliography) discovered the phenomenon of crack closure, which can help in explaining the effect of R on crack growth rates. Crack closure occurs as a consequence of crack tip plastici-

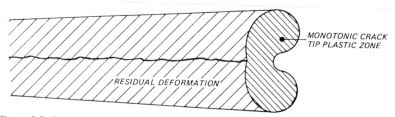

Figure 9.7. Zones of plastic deformation in the vicinity of a fatigue crack.

ty. At the tip of a growing fatigue crack each loading cycle generates a monotonic plastic zone during increased loading and a much smaller reversed plastic
zone during unloading (the reversed plastic zone is approximately one-quarter
the size of the monotonic plastic zone). Thus there is residual plastic deformation consisting of monotonically stretched material.

As the crack grows the residual plastic deformation forms a wake of monotonically stretched material along the crack flanks. This is depicted in figure 9.7
for the case of gradually increasing ΔK and hence gradually increasing plastic
zone size.

Because the residual deformation is the consequence of tensile loading the
material in the crack flanks is elongated normal to the crack surfaces and has
to be accommodated by the surrounding elastically stressed material. This is
no problem as long as the crack is open, since then the crack flanks will simply show a displacement normal to the crack surfaces. However, as the fatigue
load decreases the crack will tend to close and the residual deformation becomes
important. This will be illustrated with the help of figure 9.8, as follows:

(1) Figure 9.8.a shows the variation in the nominal stress intensity factor,
K, with applied stress, σ.

(2) Figure 9.8.b shows that as the applied stress decreases from σ_{max} the
crack tip opening angle decreases owing to elastic relaxation
of the cracked body. However, the crack surfaces are prevented from
becoming parallel because the stretched material along the flanks causes
closure before zero-load is reached. This closure results in reaction
forces and a mode I stress intensity which increases as the applied stress
decreases. This mode I stress intensity is due to crack line loading, which
is equivalent to a crack under internal pressure, see section 2.5.

(3) Figure 9.8.c shows superposition of the stress intensities due to the
applied stress and crack closure, and also the important concept of
crack opening stress, σ_{op}, which has been defined by Elber as that
value of applied stress for which the crack is just fully open. σ_{op} can
be determined experimentally from the change in compliance: crack
closure results in an increase of stiffness and a decrease in compliance.
Elber further suggested that for fatigue crack growth to occur the
crack must be fully open. Thus an effective stress intensity factor range,
ΔK_{eff}, can be defined from the stress range $\sigma_{max} - \sigma_{op}$. ΔK_{eff} is smaller
than the nominal ΔK.

It should be noted that Elber's suggestion that fatigue crack growth occurs only when the crack is fully open is not entirely correct. In fact it would be better to define ΔK_{eff} as $K_{max} - K_{min,eff}$, see figure 9.8.c. However, $K_{min,eff}$ cannot be determined experimentally.

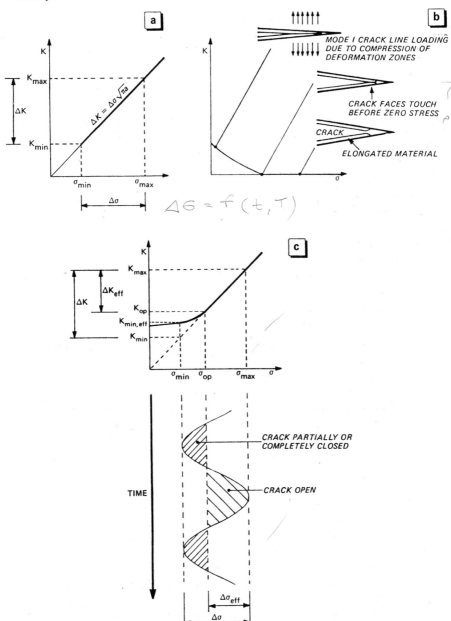

Figure 9.8. Principal of crack closure: (a) nominal K-σ plot, (b) residual deformation due to crack tip plasticity results in mode I crack line loading K values, compare section 2.5, (c) superposition of K values shows the effect of crack closure.

Higher R values result in less crack closure, i.e. ΔK_{eff} becomes more nearly equal to ΔK. Elber therefore proposed that ΔK_{eff} accounts for the effect of R on crack growth rates, so that

$$\frac{da}{dn} = f(\Delta K_{eff}) \tag{9.5}$$

cf. equation (9.4). He also obtained the empirical relationship

$$\frac{\Delta K_{eff}}{\Delta K} = U = 0.5 + 0.4\,R \tag{9.6}$$

which enabled crack growth rates for a wide range of R values to be correlated by ΔK_{eff}. This relationship, which was obtained for the aluminium alloy 2024-

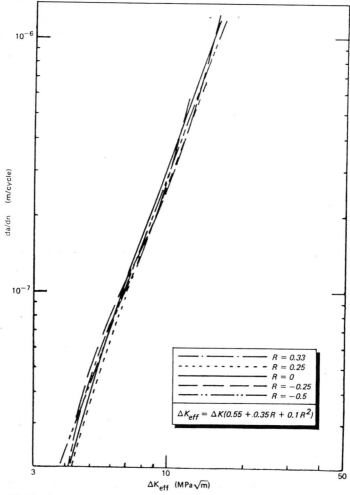

Figure 9.9. Crack growth rate data from figure 9.6 correlated by ΔK_{eff}. (Reference 4 of the bibliography.)

T3, has been modified by Schijve (reference 8 of the bibliography) as follows:

$$U = 0.55 + 0.35R + 0.1R^2. \tag{9.7}$$

The usefulness of equation (9.7) is demonstrated by figure 9.9, in which the crack growth rate data from figure 9.6 are plotted against ΔK_{eff}.

9.4. Environmental Effects

Fatigue crack growth is a complex process influenced by a number of variables besides the effective crack tip stress intensity field. In particular the environment and material microstructure can have large influences in various regions of the crack growth rate curve, figure 9.3.

Environmental effects are very important in regions I and II of the crack growth rate curve. Several types of environmental effects are schematically illustrated in figure 9.10. The first type, figure 9.10.a, is a reduction or elimination of ΔK_{th} and an overall enhancement of crack growth rates until high ΔK levels, at which the purely mechanical contribution to crack growth predominates.

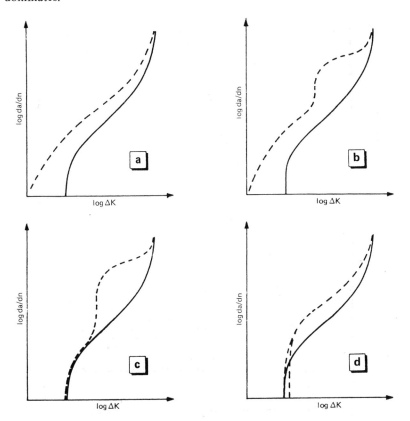

Figure 9.10. Environmental effects on the fatigue crack growth rate curve: normal environment (e.g. air) ————— ; aggressive environment – – – – –.

A second type of environmental effect, figure 9.10.b, diifers from the first in that above an intermediate ΔK level there is an additional enhancement of crack growth rates owing to monotonic, sustained load fracture during the tensile part of each stress cycle. This behaviour is typical for fatigue in liquid environments that cause stress corrosion. On the other hand a gaseous environment, usually hydrogen, that causes sustained load fracture can give the type of behaviour shown in figure 9.10.c. (Sustained load fracture is discussed as a separate topic in chapter 10.)

Finally, figure 9.10.d shows a type of environmental effect in which there is an overall enhancement of crack growth rates except near the threshold. In such cases ΔK_{th} in the aggressive environment can even be higher! This can be explained in some instances by local corrosion on the crack surfaces. The corrosion products increase the volume of material contributing to crack closure, thereby raising σ_{op} and decreasing ΔK_{eff}. This will be discussed in more detail in chapter 13, section 13.4.

Remarks

(1) Despite the last statement about corrosion products and crack closure it is not intended here to discuss possible mechanisms of environmental fatigue crack growth. This is a controversial subject, since material–environment interactions at crack tips are difficult to observe and apparently can be highly complex.

(2) Environmental effects on fatigue crack growth strongly depend on the specific material–environment combination and also on several other factors. The main ones are:

- Frequency of the fatigue stress cycle: low frequencies generally result in greater environmental effects.
- Waveform of the fatigue stress cycle: in general, higher crack growth rates occur if the increasing part of the load cycle occurs more slowly. For example, a positive sawtooth waveform ∿∿∿∿ results in higher environmental fatigue crack growth rates than a negative sawtooth waveform ∿∿∿∿ .
- Temperature: environmental effects are usually greater at higher temperatures.
- A very high growth rate (high ΔK) diminishes the importance of the environmental effect (K_{max} during fatigue nears K_c, see figure 9.3).

(3) The unusual shapes of the environmental fatigue crack growth rate curves in figures 9.10.b and 9.10.c deserve comment. At intermediate ΔK levels the overall crack growth rate suddenly increases greatly, followed by crack growth rates with a relatively weak dependence on ΔK. This is a consequence of major contribution to the overall crack growth by sustained load fracture. As will be discussed in chapter 10, sustained load fracture crack growth rates can be almost constant over a wide range of stress intensity.

9.5. Prediction of Fatigue Crack Growth Under Constant Amplitude Loading

The main purpose of crack growth prediction is to construct a crack growth curve, a versus n, like that shown schematically in figure 9.2. For constant amplitude loading this is a fairly straightforward procedure of integrating da/dn−ΔK curves.

It is often possible to use "crack growth laws" like equations 9.1−9.3. These are useful because they are closed form expressions that can be integrated analytically. For example, the crack growth rate data for BS 4360 structural steel shown in figure 9.4 can be expressed as

$$\frac{da}{dn} = 5.01 \times 10^{-12} \, (\Delta K)^{3.1} \;\; m/cycle \tag{9.8}$$

if ΔK is in MPa√m. For a wide plate one can substitute $\Delta K = \Delta \sigma \sqrt{\pi a}$ in equation (9.6), giving

$$da = 5.01 \times 10^{-12} \, (\Delta \sigma \sqrt{\pi a})^{3.1} \, dn$$

or

$$dn = \frac{da}{5.01 \times 10^{-12} \, (\Delta \sigma \sqrt{\pi a})^{3.1}}$$

and

$$n = \frac{10^{12}}{5.01 \, (\Delta \sigma \sqrt{\pi})^{3.1}} \int_{a_d}^{a_{cr}} (a)^{-1.55} \, da. \tag{9.9}$$

(Note that the crack length a has to be in metres (m) because of the units for ΔK.)

However, if finite width correction factors or more complicated crack growth laws are necessary it becomes difficult to integrate crack growth rate data analytically. Instead a numerical approach must be used. This can be done in the following way:

- choose a suitable increment of crack growth, $\Delta a_i = a_{i+1} - a_i$
- calculate ΔK for the crack length corresponding to the mean of the crack growth increment, i.e. $(a_{i+1} + a_i)/2$
- determine da/dn for this value of ΔK
- calculate Δn_i from $\Delta a_i/(da/dn)_i$
- repeat the previous steps over the required range of crack growth and sum the values of Δn_i.

Thus it is possible to predict crack growth from any type of da/dn−ΔK curve as long as the relation between crack length, a, and ΔK is known for the structure or component under consideration. Herein lies the difficulty: in practice it may be difficult to estimate ΔK owing to the complex geometry of the crack (e.g. semi elliptical surface flaws and quarter elliptical corner cracks) and to the occurrence of load shedding. In built-up structures cracked elements will shed load to uncracked elements because cracking causes a decrease in stiffness and the displacements in each element are mutually constrained in a kind of "fixed grip" condition (see also section 4.2).

9.6. Fatigue Crack Growth Under Variable Amplitude Loading

Although constant amplitude loading does occur in practice (e.g. pressuri-zation cycles in transport aircraft cabins, rotating bending stresses in generators, thermal stress cycles in pressure vessels) the vast majority of dynamically loaded structures actually experience variable amplitude loading that is often fairly random. An example is given in figure 9.11.

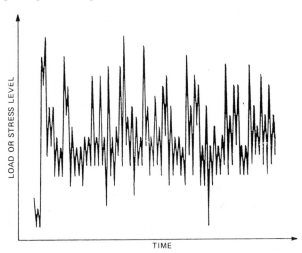

Figure 9.11. Example of a fairly random cyclic load history (part of a flight for a tacti-cal aircraft).

An important consequence of variable amplitude loading is that ΔK need not increase gradually with increasing crack length. For example, large and small stress ranges follow each other directly in figure 9.11 and so the instan-taneous ΔK will vary greatly. This variation results in load interaction effects which may strongly influence fatigue crack growth rates. The nature of such interaction effects is best demonstrated by some simple examples.

Simple Examples of Variable Amplitude Loading

The simplest type of variable amplitude loading is the occurrence of occa-sional peak loads in an otherwise constant amplitude loading history. Figure 9.12 shows two such simple variable amplitude loading histories with their effects on fatigue crack growth as compared to constant amplitude loading. There is a profound effect on crack growth owing to the occurrence of positive peak stresses only (curve C). However, the effect is much less when positive peak stresses are immediately followed by negative peak stresses (curve B). The explanation for these effects is as follows:

(1) Positive peak stresses (curve C). Each peak stress opens up the crack tip much more than the normal maximum stress, but also creates a larger plastic zone ahead of the crack. There are three consequences when normal load cycling is resumed. Initially the amount of crack

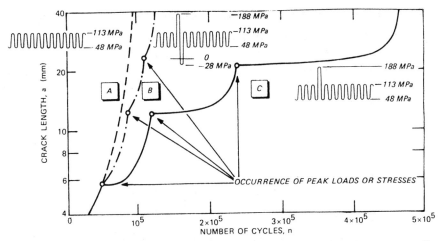

Figure 9.12. Crack growth curves under constant amplitude loading and constant amplitude loading + occasional peak loads for a centre cracked panel of aluminium
alloy 2024-T3 (reference 9 of the bibliography).

closure is reduced (the crack may be fully open at minimum stress
with $\Delta K_{eff} = \Delta K$) and the crack will grow somewhat faster than before the peak stress occurred. However, the crack soon grows into the
large plastic zone and encounters high residual compressive stresses
because this plastic zone must be accommodated by the surrounding
elastically stressed material. Also a wake of enhanced residual deformation forms behind the crack tip and causes an increase in crack
closure.

The overall result is a significant retardation of crack growth. This is
often called delayed retardation, owing to the initial acceleration of
crack growth following the peak stress.

(2) Positive + negative peak stresses (curve B). The negative peak stress
reverses most of the tensile plastic deformation due to the positive
peak stress and the amount of crack growth retardation is greatly dim-
inished.

Since residual stresses and residual plastic deformation in the vicinities of
crack tips are responsible for interaction effects, the magnitude of these effects
will depend on the ratios of the peak stresses to the normal maximum stresses,
the material yield strength and strain hardening characteristics, and the stress
state (plane strain or plane stress). The larger plastic zones in plane stress
result in much greater interaction effects.

More Complex Variable Amplitude Loading

For the purpose of this course variable amplitude loading can be placed in
two categories:
- stationary variable amplitude loading
- non-stationary variable amplitude loading.

184

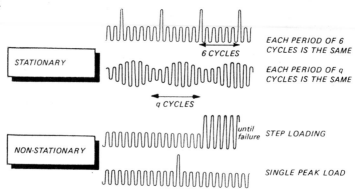

Figure 9.13. Examples of stationary and non-stationary variable amplitude loading.

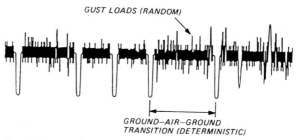

Figure 9.14. Example of load history for a transport aircraft wing.

Examples of both categories are shown in figure 9.13. This shows that repetitions of the same sequence of load cycles are examples of stationary variable amplitude loading. These sequences of load cycles are deterministic, i.e. non-random. However, a random load sequence may also be classified as stationary if its statistical description is constant, i.e. independent of time.

In practice the loads are often a mixture of deterministic and random loads. For example, a transport aircraft experiences deterministic ground-air-ground transitions (one cycle per flight) and random gust loads (many cycles per flight), figure 9.14. Provided that the statistical description of this mixture of loads is invariant with time, then such a load sequence also may be classified as stationary variable amplitude loading.

In the previous part of this section it was shown that positive peak loads can cause crack growth retardation. In turn this retardation may result in crack growth no longer being a regular (i.e. stationary) process. To illustrate this, consider the loading histories shown in figure 9.15. The first load history has peak loads with a long recurrence period. Such a load history can result in large discontinuities in the crack growth curve. Curve C in figure 9.12 is a good example. These discontinuities are pronounced because the recurrence period is sufficiently long that each retardation is over before the next peak load occurs.

On the other hand, figure 9.15 shows that the second load history, which has peak loads with a short recurrence period, results in an almost regular crack growth curve. Retardations still occur, but they are superimposed on each other: this is, in fact, highly effective in slowing down the overall crack growth.

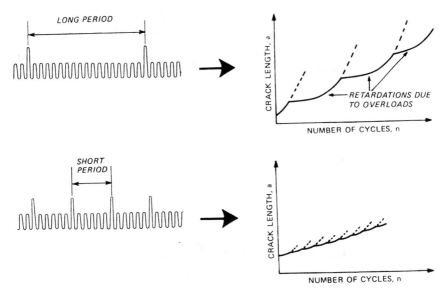

Figure 9.15. Effect of peak load recurrence period on the crack growth curve.

It is thus clear that the recurrence periods of peak loads are important for crack growth under variable amplitude loading. Long recurrence periods will disturb the regularity of the crack growth process and it becomes non-stationary. Short recurrence periods result in almost regular and stationary crack growth behaviour. These differences have consequences for the choice of methods to predict crack growth under variable amplitude loading.

9.7. Prediction of Fatigue Crack Growth Under Variable Amplitude Loading

Owing to load interaction effects it cannot be assumed that an increment of crack growth, Δa_i, in cycle i of a variable amplitude load history is directly related to ΔK_i. Thus a straightforward summation procedure using constant amplitude crack growth rate data (as in section 9.5) will not give proper results. Nevertheless such a summation is sometimes done. The results are nearly always conservative, often extremely so, since the most important consequence of load interaction effects is crack growth retardation.

An empirical way of accounting for load interaction effects is to determine experimentally whether crack growth rates under various kinds of variable amplitude loading can be correlated by characteristic stress intensity factors, and if so to numerically integrate the crack growth rate curves in the same way as for constant amplitude loading. This method is subject to several limitations, as will be discussed.

Besides this characteristic K method a number of crack growth models have been developed since about 1970 to try and account for load interaction effects and thereby enable predictions of crack growth curves. These models are of two kinds:

(1) Models based on crack tip plasticity ("first generation" models).
(2) Models based on crack closure ("second generation" models).

A concise review of these models will be made.

The Characteristic K Method

For some types of variable amplitude load history fatigue crack growth is a regular process because peak loads have either a short recurrence period or only minor effects on crack extension in subsequent load excursions. In such cases it is often possible to correlate crack growth rates by a characteristic stress intensity factor. One possibility is the root mean value of the stress intensity factor range, ΔK_{rm}. The general expression for ΔK_{rm} is

$$\Delta K_{rm} = \sqrt[m]{\frac{\Sigma (\Delta K_i)^m n_i}{\Sigma n_i}} \qquad (9.10)$$

where n_i is the number of load amplitudes corresponding to ΔK_i and m is the slope of the constant amplitude da/dn versus ΔK plot, i.e. the exponent in the Paris equation, equation (9.1).

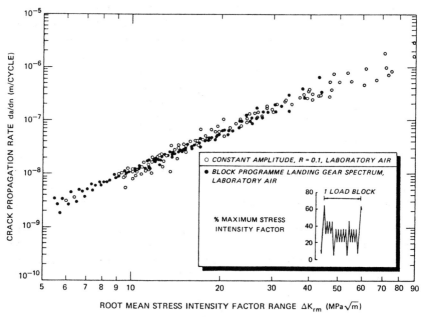

Figure 9.16. Crack growth rates in three ultrahigh strength steels correlated by ΔK_{rm}.

The main problem in the use of equation (9.10) is the derivation of ΔK_i values. To do this properly it is necessary to account for crack closure and follow a rational procedure of cycle counting. A full discussion of this problem is given in reference 10 of the bibliography, from which the example in figure 9.16 is taken. This figure shows that ΔK_{rm} correlates regularly retarded crack growth under block programma loading with constant amplitude crack growth. The reason is that correct derivation of ΔK_{rm} for block programme loading results in a much smaller value than the ΔK_{rm} $(= \Delta K_{eff})$ for constant amplitude loading, and this succesfully accounts for the shift in the block programme crack growth rate curve owing to retardation.

A special case is narrow band random loading of steels, for which m was assumed to be 2 and no account was taken of crack closure, reference 11. When m = 2 the stress intensity factor range becomes the root mean square value, ΔK_{rms}. Since the root mean square is a parameter characterizing a stationary random process it was considered that ΔK_{rms} should correlate crack growth under stationary (narrow band) random loading. In fact, good correlations were obtained. However, such correlations are very empirical and are useful for predicting crack growth only for load histories very similar to those for which the correlations were made.

Crack Growth Models Based on Crack Tip Plasticity

These models assume that after a peak load there will be an interaction effect as long as the crack tip plastic zones for subsequent load cycles are within the plastic zone due to the peak load. The consequences of this assumption will be illustrated using figure 9.17 and a well known crack growth model introduced by Wheeler (reference 12 of the bibliography).

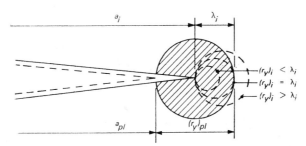

Figure 9.17. Possible crack tip plastic zone sizes due to a load cycle following a peak load.

Consider a crack that has grown to length a_i after a peak load at length a_{pl}. The plastic zone size due to the peak load is $(r_y)_{pl}$. There are three possible situations for the plastic zone size $(r_y)_i$ due to load cycle i (which follows the peak load):

$$(r_y)_i < \lambda_i \quad \text{or} \quad \frac{(r_y)_i}{\lambda_i} < 1$$

$$(r_y)_i = \lambda_i \quad \text{or} \quad \frac{(r_y)_i}{\lambda_i} = 1 \qquad (9.11)$$

$$(r_y)_i > \lambda_i \quad \text{or} \quad \frac{(r_y)_i}{\lambda_i} > 1.$$

Wheeler defined a crack growth retardation factor, ϕ, as follows:

$$\phi_i = \left[\frac{(r_y)_i}{\lambda_i}\right]^m \qquad \text{for} \qquad \frac{(r_y)_i}{\lambda_i} < 1$$

$$\phi_i = 1 \qquad \text{for} \qquad \frac{(r_y)_i}{\lambda_i} \geqslant 1. \tag{9.12}$$

The plastic zone sizes $(r_y)_{pl}$ and $(r_y)_i$ are given by

$$r_y = \frac{1}{C\pi}\left[\frac{K_{max}}{\sigma_{ys}}\right]^2 \tag{9.13}$$

and the increment of crack growth, Δa_i, occurring during load cycle i is

$$\Delta a_i = \phi_i\left(\frac{da}{dn}\right)_{ca} \tag{9.14}$$

where $(da/dn)_{ca}$ is the constant amplitude crack growth rate appropriate to the stress intensity factor range, ΔK_i, and stress ratio, R_i, of load cycle i. Note that as soon as $\phi_i = 1$ there is no longer any retardation.

The constants m and C in equations (9.12) and (9.13) are empirical. They have to be found by fitting crack growth predictions to test data. For this purpose a fairly simple variable amplitude load history is used. Once m and C are obtained the crack growth curve for a complex variable amplitude load history is predicted using a step by step counting method for each crack length a_i:

- calculate $(r_y)_{pl}$ and $(r_y)_i$ according to equation (9.13)
- calculate λ_i ($= a_{pl} + (r_y)_{pl} - a_i$)
- calculate ϕ_i according to equation (9.12) or set $\phi_i = 1$ if $(r_y)_i \geqslant \lambda_i$
- calculate Δa_i according to equation (9.14)
- determine $a_{i+1} = a_i + \Delta a_i$
- repeat the previous steps for each load cycle.

Note that this procedure is basically different from the numerical integration of constant amplitude data mentioned in section 9.5. For constant amplitude crack growth prediction a crack length increment is taken and the cycles are summed. For variable amplitude crack growth prediction each cycle is taken and the crack length increments are summed.

Crack growth models like that of Wheeler have some success in predicting crack growth under variable amplitude loading. However, the models rely completely on empirical adjustments (m, C) and do not incorporate effects such as delayed retardation or the counteracting effect of negative peak loads discussed earlier in this section. These limitations make the models unreliable for truly predicting crack growth. In other words they are useful mainly for curve fitting and interpolation of test results, but not for extrapolation to significantly different load histories.

Crack Growth Models Based on Crack Closure

More recent crack growth models incorporate crack closure. This enables delayed retardation and the effect of negative peak loads to be accounted for by the variation in σ_{op} and hence ΔK_{eff}. The main problem is to determine the correct variation in σ_{op} during variable amplitude loading. Discussion of specific solutions to this problem is beyond the scope of this course. Interested readers may consult references 13 and 14 of the bibliography.

Once the variation in σ_{op} has been determined the procedure for estimating the crack growth curve is as follows:

- calculate $(\sigma_{op})_i$ for load cycle i as a function of R and previous positive and negative load cycles (e.g. by the method in reference 13)
- determine $(\Delta\sigma_{eff})_i = (\sigma_{max})_i - (\sigma_{op})_i$ and hence $(\Delta K_{eff})_i$
- calculate $\Delta a_i = (da/dn)_i = f[(\Delta K_{eff})_i]$: for this step constant amplitude crack growth rate data correlated by ΔK_{eff} are necessary, e.g. figure 9.9
- determine $a_{i+1} = a_i + \Delta a_i$
- repeat the previous steps for each load cycle.

The relatively complex way in which σ_{op} must be determined for each load cycle and the step by step counting method of generating the crack growth curve requires a large computer programme and long running times. However, the results are very promising in terms of both interpolation of test data and truly predicting crack growth.

Recently Elber (reference 15 of the bibliography) proposed using a constant σ_{op} determined from constant amplitude testing of the same material. This would eliminate the tedious calculation of σ_{op} for each load cycle, but at the same time would limit the use of crack closure based models to variable amplitude load histories that result in fairly regular crack growth.

9.8. Fatigue Crack Growth from Notches: the Short Crack Problem

Most fatigue failures originate at some kind of stress concentration or notch. Owing to the presence of a notch crack growth often begins under conditions of local plasticity. The crack then proceeds through the elastic stress-strain field of the notch before it reaches the bulk stress-strain field. This situation is depicted schematically in figure 9.18.

When the crack is propagating in the notch plastic field it is incorrect to use LEFM to characterize crack growth. This is demonstrated by the fact that crack growth rates are much higher than those expected on the basis of the nominal ΔK values. However, even in the notch elastic field the crack growth rate data for short cracks are not correlated by ΔK.

There are three kinds of deviation from the normal da/dn–ΔK curve, figure 9.19. For notches with both plastic and elastic stress-strain fields the crack growth rates are initially high but decrease with increasing crack length (and nominal ΔK) because growth is controlled by the plastic strain range, which diminishes rapidly. In the notch elastic field crack growth rates may or may not correlate with the normal da/dn–ΔK curve. This depends on the crack length: short cracks tend to grow faster than long cracks at the same ΔK.

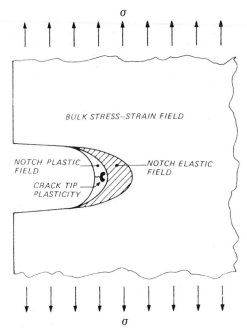

Figure 9.18. Growth of a small fatigue crack at a notch and the associated elastic and plastic stress-strain fields.

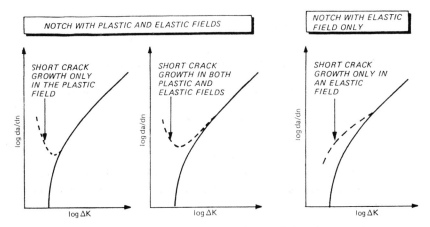

Figure 9.19. Schematic of anomalous crack growth rates for cracks at notches.

A major reason for short cracks in the notch elastic field to grow faster than long cracks is the difference in residual plastic deformation in the wakes of the cracks. Figure 9.20 compares short and long cracks with the same ΔK. The lesser amount of residual deformation for the short crack results in less crack closure, a higher ΔK_{eff} and hence a higher crack growth rate.

Several attempts have been made to account for the faster growth of short cracks at notches, e.g. references 16–18 of the bibliography. They all provide "corrections" to the physical crack length. Some examples are given here.

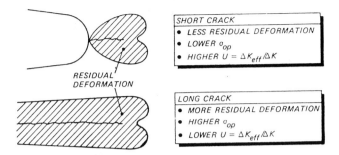

Figure 9.20. Differing plasticity in the wakes of short and long cracks with the same ΔK.

(1) Crack within the notch elastic field. A fairly successful correction is that of El Haddad et al.:

$$\Delta K = k' \Delta\sigma\sqrt{\pi(a + l_0)} \qquad (9.15)$$

where k' is the elastic concentration factor accounting for the increase in crack tip stress owing to the notch and l_0 is supposed to be a material constant. This constant is evaluated from the limiting condition where the physical crack size approaches zero. Then $l_0 \gg a$; $k'\Delta\sigma \to \Delta\sigma_e$, the fatigue limit of the material (see figure 9.1); $\Delta K \to \Delta K_{th}$; and equation (9.15) can be rewritten as

$$\Delta K_{th} = \Delta\sigma_e\sqrt{\pi l_0} \quad\text{or}\quad l_0 = \frac{1}{\pi}\left[\frac{\Delta K_{th}}{\Delta\sigma_e}\right]^2. \qquad (9.16)$$

Another useful correction is due to Smith and Miller:

$$\text{for } a < 0.13\sqrt{D\rho}, \quad \Delta K = \Delta\sigma\sqrt{\pi a}\,[1 + 7.69\sqrt{D/\rho}\,]^{\frac{1}{2}}$$
$$\text{for } a > 0.13\sqrt{D\rho}, \quad \Delta K = \Delta\sigma\sqrt{\pi(a + D)} \qquad (9.17)$$

where ρ is the actual notch root radius and D is the depth of an elliptical notch with the same root radius.

(2) Crack within the notch plastic field. For this case El Haddad et al. proposed using either ΔJ for the corrected crack length $(a + l_0)$ or else a strain-based intensity factor, ΔK_ϵ, defined as

$$\Delta K_\epsilon = E\,\Delta\epsilon\sqrt{\pi(a + l_0)} \qquad (9.18)$$

where $\Delta\epsilon$ is the local plastic strain. Equation (9.18) appears simple, but $\Delta\epsilon$ is not easy to estimate. The same is true of ΔJ. Interested readers should consult references 17 and 18 of the bibliography.

The problem of short crack growth at notches is important because most fatigue failures originate at some kind of stress concentration, as stated earlier, and the period of crack growth through the notch stress field may represent a major part of the total life. In such cases a proper description of short crack growth is essential to accurate predictions of the fatigue life.

9.9. Bibliography

1. Johnson, R., Bretherton, I., Tomkins, B., Scott, P.M. and Silvester, D.R.V., *The Effects of Sea Water Corrosion on Fatigue Crack Propagation in Structural Steel*, European Offshore Steels Research Seminar, The Welding Institute, pp. 387–414 (1980): Cambridge.

2. Crooker, T.W., *The Role of Fracture Toughness in Low Cycle Fatigue-Crack Propagation for High Strength Alloys*, Engineering Fracture Mechanics, Vol. 5, pp. 35–43 (1973).

3. Van der Veen, J.H., van der Wekken, C.J. and Ewalds, H.L., *Fatigue of Structural Steels in Various Environments*, Delft University of Technology, Department of Metallurgy Report, October 1978.

4. Ewalds, H.L., van Doorn, F.C. and Sloof, W.G., *Influence of the Environment and Specimen Thickness on Fatigue Crack Growth Data Correlation by means of Elber Type Equations*, Corrosion Fatigue, ASTM STP 801, pp. 115–134 (1983): Philadelphia.

5. Barsom, J.M., Imhof, E.J. and Rolfe, S.T., *Fatigue-Crack Propagation in High-Strength Steels*, Engineering Fracture Mechanics, Vol. 2, pp. 301–317 (1971).

6. De Vries, M.I. and Staal, H.U., *Fatigue-Crack Propagation of Type 304 Stainless Steel at Room Temperature and 550°C*, Energy Research Centre Report #RCN 222 (1975): Petten, The Netherlands.

7. Elber, W., *The Significance of Fatigue Crack Closure*, Damage Tolerance in Aircraft Structures, ASTM STP 486, pp. 230–247 (1971): Philadelphia.

8. Schijve, J., *The Stress Ratio Effect on Fatigue Crack Growth in 2024-T3 Alclad and the Relation to Crack Closure*, Delft University of Technology Department of Aerospace Engineering, Memorandum M-336, August 1979.

9. Schijve, J. and Broek, D., *Crack Propagation: the Results of a Test Programme Based on a Gust Spectrum with Variable Amplitude Loading*, Aircraft Engineering, Vol. 34, pp. 314–316 (1962).

10. Wanhill, R.J.H., *Fatigue Fracture in Steel Landing Gear Components*, National Aerospace Laboratory Technical Report 84117, October 1984.

11. Barson, J.M., *Fatigue Crack Growth Under Variable-Amplitude Loading in Various Bridge Steels*, Fatigue Crack Growth Under Spectrum Loads, ASTM STP 595, pp. 217–235 (1976): Philadelphia.

12. Wheeler, O.E., *Spectrum Loading and Crack Growth*, Journal of Basic Engineering, Transactions ASME, Vol. 94, pp. 181–186 (1972).

13. De Koning, A.U., *A Simple Crack Closure Model for Prediction of Fatigue Crack Growth Rates under Variable-Amplitude Loading*, Fracture Mechanics, ASTM STP 743, pp. 63–85 (1981): Philadelphia.

14. Newman, J.C., Jr., *Prediction of Fatigue Crack Growth under Variable-Amplitude and Spectrum Loading Using a Closure Model*, Design of Fatigue and Fracture Resistant Structures, ASTM STP 761, pp. 255–277 (1982): Philadelphia.

15. Elber, W., *Equivalent Constant-Amplitude Concept for Crack Growth under Spectrum Loading*, Fatigue Crack Growth under Spectrum Loads, ASTM STP 595, pp. 236–250 (1976): Philadelphia.

16. Smith, R.A. and Miller, K.J., *Prediction of Fatigue Regimes in Notched Components*, International Journal of Mechanical Science, Vol. 20, pp. 201–206 (1978).

17. El Haddad, M.H., Dowling, N.E., Topper, T.H. and Smith, K.N., *J Integral Applications for Short Fatigue Cracks at Notches*, International Journal of Fracture, Vol. 16, pp. 15–30 (1980).

18. El Haddad, M.H., Topper, T.H. and Topper, T.N., *Fatigue Life Predictions of Smooth and Notched Specimens Based on Fracture Mechanics*, Journal of Engineering Materials and Technology, Vol. 103, pp. 91–96 (1981).

10. SUSTAINED LOAD FRACTURE

10.1. Introduction

Sustained load fracture is a general term for time-dependent crack growth under loading often well below that normally required to cause failure in a tensile or fracture toughness test. Examples of sustained load fracture are:

- creep and creep crack growth
- stress corrosion cracking
- cracking due to embrittlement by internal or external (gaseous) hydrogen
- liquid metal embrittlement.

Creep deformation and cracking constitute a widespread and very important practical problem, particularly in the power generating industry and aircraft gas turbines. However, a good and universally accepted description of creep cracking by fracture mechanics, whether LEFM or EPFM, is as yet unavailable.

The other types of sustained load fracture follow basically similar trends in terms of fracture mechanics. Thus, excluding creep, it is possible to discuss the application of fracture mechanics to sustained load fracture in a general way. As with fatigue crack growth the use of fracture mechanics is mainly limited to LEFM methods, specifically the stress intensity (K) approach.

The use of K to describe sustained load fracture is based mainly on procedures similar to those for fracture toughness testing, including the use of more or less standard specimens with fatigue precracks under nominally plane strain conditions. However, full plane strain is not strictly necessary if it is only required to test specimens with a thickness representative of that for a narrow section component in service.

Experimental methods for fracture mechanics evaluation of sustained load fracture fall into two categories:

(1) Time-to-failure (TTF) tests on precracked specimens.
(2) Crack growth rate testing.

These methods will be discussed in sections 10.2 and 10.3 respectively, and the experimental problems that arise are dealt with in section 10.4.

The way in which crack growth rate data could be used to predict failure of a structural component in service is treated in section 10.5. However, there are many difficulties that require careful evaluation of the practical significance of test data. This is the subject of section 10.6, which closes the chapter.

10.2. Time-To-Failure (TTF) Tests

For a long time the study of sustained load fracture (principally stress corrosion cracking) relied solely upon TTF tests on smooth specimens. Such tests are still useful, but it is now recognised that fracture mechanics based tests provide an essential supplement. The most striking example concerns titanium alloys, which were thought to be immune to stress corrosion in aqueous solutions until Brown in 1966 tested fatigue precracked cantilever

194

beam specimens with disastrous results (reference 1 of the bibliography to this chapter).

Precracked specimens for TTF tests are configured such that a constant load results in increasing stress intensity with increasing crack length. The specimens are loaded to various initial stress intensity levels, K_{I_i}, and the time to failure is recorded. A representative TTF plot is shown in figure 10.1.

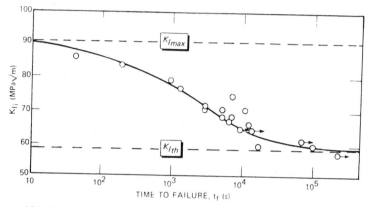

Figure 10.1. Sustained load testing of precracked titanium alloy Ti-6Al-4V specimens in normal air.

The plot shows that two quantities, $K_{I_{max}}$ and $K_{I_{th}}$, can be determined. $K_{I_{max}}$ represents the maximum load carrying ability and will generally be equal to either K_{Ic} or K_Q, the valid or invalid fracture toughness values. $K_{I_{th}}$ is the threshold stress intensity, below which there is virtually no crack growth. For stress corrosion cracking the threshold is customarily referred to as $K_{I_{scc}}$.

In some cases $K_{I_{th}}$ may be a true threshold. In general, however, $K_{I_{th}}$ should not be considered a material property. There are two reasons for this. First, the testing time required to establish $K_{I_{th}}$ may be extremely long (years rather than months). Secondly, some materials (notably steels) exhibit long incubation periods before sustained load fracture commences from the fatigue precrack, and the incubation periods increase with decreasing K_{I_i}.

Figure 10.2 shows schematically that in the regime between $K_{I_{th}}$ and $K_{I_{max}}$

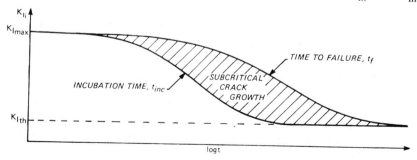

Figure 10.2. Schematic TTF plot.

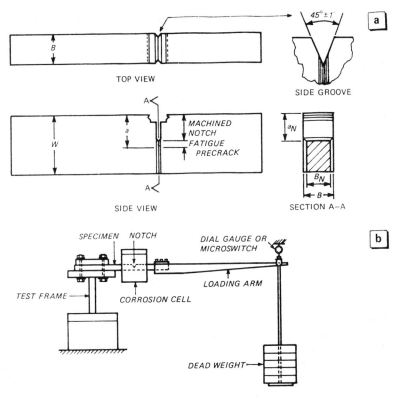

Figure 10.3. (a) Fatigue precracked cantilever beam specimen and
(b) loading arrangement for stress corrosion testing.

the time to failure, t_f, generally comprises an incubation time, t_{inc}, and a period of subcritical crack growth. The incubation time depends on the material, environment, K_{I_i} and also on the previous loading history (if any) of the specimen. The subcritical crack growth period depends on the specimen configuration, type of loading, amount of crack growth before $K_{I_{max}}$ is reached, and the kinetics of crack growth due to the material-environment interaction.

As stated earlier, precracked specimens for TTF tests are such that a constant load results in increasing stress intensity with increasing crack length. A commonly used specimen is the fatigue precracked cantilever beam specimen developed by Brown and coworkers at the U.S. Naval Research Laboratory. The specimen and loading arrangement are illustrated in figure 10.3.

The cantilever beam specimen usually has side grooves reducing the nominal width from B to B_N. Side grooves are beneficial in obtaining a more even stress distribution through the thickness, since they prevent the occurrence of a plane stress state at the specimen side surfaces. The results are a more uniform crack front and prevention of out-of-plane crack deviation for some specimen types.

Stress intensity factors for cantilever beam specimens can be calculated according to:

$$K_I = \frac{6M}{(B \cdot B_N)^{\frac{1}{2}}(W - a)^{\frac{3}{2}}} \cdot f(\frac{a}{W}),$$

(10.1)

where M is the bending moment, W is the beam depth, and a is the total crack length, see figure 10.3. The geometry factor, f(a/W), is given in the following table.

$(\frac{a}{W})$	$f(\frac{a}{W})$
0.05	0.36
0.10	0.49
0.20	0.60
0.30	0.66
0.40	0.69
0.50	0.72
$\geqslant 0.60$	0.73

10.3. Crack Growth Rate Testing

The general features of the dependence of sustained load crack growth on K_I were already mentioned in section 1.9 (figure 1.12) and are shown again in figure 10.4. The crack growth curve consists of three regions. In regions I and III the growth rate da/dt strongly depends on the stress intensity, but in region II the crack growth rate is virtually independent of stress intensity. Regions I and II are the most characteristic, although region II is sometimes a rounded hump rather than a plateau. Region III is often not observed, owing to an abrupt transition from region II to fast fracture. Note that $K_{I_{th}}$ and $K_{I_{max}}$ can be determined, at least in principle, from crack growth rate tests.

Many types of specimen have been proposed for studying sustained load crack growth. Most fall into either the increasing K (constant load) or decreasing K (constant displacement) categories: compare also with section 5.4.

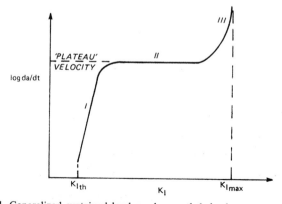

Figure 10.4. Generalized sustained load crack growth behaviour.

Advantages of decreasing K specimens are:

- Fatigue precracking from a sharp notch is not always necessary, since initial mechanical crack growth ('pop-in') is stable and usually does not influence subsequent sustained load fracture.
- Virtually the entire crack growth curve can be obtained from one specimen.
- They can be self-stressed (e.g. figure 10.5) and therefore portable: for example, they can be exposed outdoors.
- Steady state conditions for crack growth and arrest at $K_{I_{th}}$ are more readily achieved.

A disadvantage of using decreasing K specimens is the occurrence of corrosion product wedging for some material-environment combinations. For a decreasing K specimen the initial displacement is fixed, so that the crack tip tends to narrow during propagation. Corrosion products may form a wedge between the crack surfaces and lead to a higher crack growth rate at a given nominal K_I and to apparently lower $K_{I_{th}}$ values. These effects are very difficult to assess quantitatively.

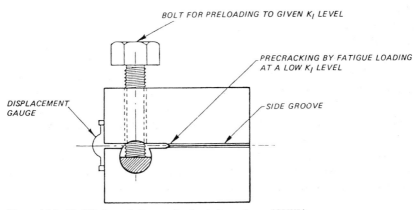

Figure 10.5. Modified crack line wedge-loaded specimen (CLWL).

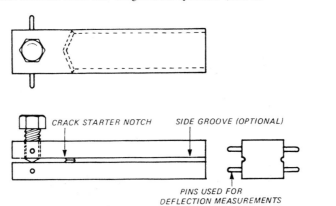

Figure 10.6. Bolt loaded double cantilever beam specimen (DCB).

Two types of commonly used decreasing K specimens, the modified crack line wedge-loaded (CLWL) and the double cantilever beam (DCB) specimens, are shown in figures 10.5 and 10.6. These types of specimens (though not self-stressed) have already been mentioned in sections 5.4 and 4.4 respectively. Both types require determination of the elastic compliance in order to calculate stress intensity factors, and the expressions for K_I are — unfortunately — cumbersome, equations (5.2) and (4.25). Some more guidance for the calculation of stress intensity factors in CLWL specimens is given in reference [2].

There is a third category of specimens, namely those giving a constant K under constant load. Such specimens have not been so widely used, but for detailed studies of sustained load fracture they provide a valuable adjunct to increasing and decreasing K tests. An example is the tapered double cantilever beam (TDCB) specimen, figure 10.7. As mentioned in section 4.4, for this specimen $(3a^2 + h^2)/h^3$ is constant. This results in a linear increase in compliance with crack length, i.e. dC/da is constant. For a given load P the stress intensity factor under plane strain conditions is given by

$$K_I = P \left[\frac{E}{2B_N(1 - \nu^2)} \cdot \frac{dC}{da} \right]^{\frac{1}{2}}$$

(10.2)

Hence K_I is constant when dC/da is constant. In practice the contour in figure 10.7 is often approximated by a straight line in order to facilitate specimen manufacture.

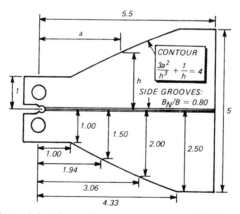

Figure 10.7. Tapered double cantilever beam specimen (TDCB).

Difference in Behaviour for Increasing and Decreasing K Specimens

A schematic of the difference in behaviour of increasing K (constant load) and decreasing K (constant displacement) specimens is given in figure 10.8.

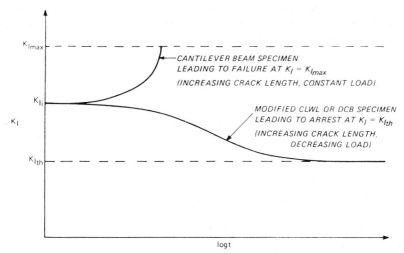

Figure 10.8. Difference in crack growth behaviour for increasing K (cantilever beam) and decreasing K (modified CLWL or DCB) specimens.

Note that the ordinate is now K_I, not K_{I_i} as in the case of TTF plots.

10.4. Experimental Problems

There are a number of experimental difficulties associated with sustained load fracture testing, including

- the incubation time, t_{inc}
- non-steady state crack growth
- effect of precrack morphology
- influence of sustained load crack morphology
- curved crack fronts
- corrosion product wedging.

Incubation Time, t_{inc}

As mentioned already in section 10.2, the incubation time, t_{inc}, depends on K_{I_i} and the previous loading history of the specimen. The loading history stems from the fatigue precracking process and can include effects of plastic deformation. Exposure to an aggressive environment before sustained loading can also affect t_{inc}. Furthermore, t_{inc} − or rather: the apparent incubation time − strongly depends on the methods of detecting crack growth, e.g. visual inspection, compliance measurements or acoustic emission.

Non-Steady State Crack Growth

A high K_{I_i} applied to a decreasing K specimen or a rapid increase in stress intensity for an increasing K specimen can lead to crack growth rates initially lower than those predicted by the $da/dt - K_I$ diagram. This non-steady state crack growth is caused by changes in the electrochemical reactions at the crack tip. (Further discussion of this interesting problem is beyond the scope of this course.)

Effect of Precrack Morphology

Stress corrosion tests on aluminium alloys with DCB specimens have shown that mechanical 'pop-in' precracks result in displacement of the crack growth curve (particularly region I, figure 10.4) to higher K_I values as compared to fatigue precracks, reference 3 of the bibliography. This was explained by the fact that fatigue precracks generate relatively planar stress corrosion cracks with uniform crack fronts, whereas pop-in cracks, being more irregular, lead to similarly irregular stress corrosion cracks which require more driving force.

Influence of Sustained Load Crack Morphology

The morphology of the sustained load crack front is especially important for tests to determine $K_{I_{th}}$. Crack tip blunting and/or microbranching in decreasing K tests lead to apparent $K_{I_{th}}$ values higher than those obtained from increasing K tests, where sustained load crack initiation occurs from a relatively sharp and unbranched fatigue precrack.

Curved Crack Fronts

Curved crack fronts often occur during sustained load fracture testing and become more pronounced as the specimen thickness is decreased. In general the reason for this curvature is the change in stress state from plane strain in the specimen interior to plane stress at the surface. Side grooves, as in figure 10.3, are helpful in reducing curvature, which is a nuisance primarily because it leads to errors in using surface crack lengths to calculate crack growth rates.

Corrosion Product Wedging

Corrosion product wedging has been mentioned already in section 10.3 as a disadvantage of using decreasing K specimens, since the wedging action leads to higher crack growth rates at a nominal K_I and to apparently lower $K_{I_{th}}$ values. To determine whether corrosion product wedging has influenced crack growth the specimen can be unloaded after testing and the deflection at the load line can be remeasured and compared to that at the beginning of the test. If the deflections are nearly the same, no substantial amount of corrosion products has accumulated in the crack and therefore wedging has not occurred.

10.5. Method of Predicting Failure of a Structural Component

As shown in figure 10.2 the time to failure, t_f, generally comprises an incubation time, t_{inc}, and a period of subcritical crack growth. Therefore in the first instance the prediction of failure of a structural component is a twofold problem:

    (1) Prediction of t_{inc}.
    (2) Prediction of the crack growth period.

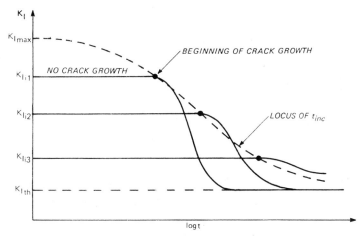

Figure 10.9. Schematic determination of t_{inc} using decreasing K specimens (reference 4).

As a reminder, the incubation time depends on the material, environment, initial stress intensity (K_{I_i}) and prior loading history. A convenient way of determining t_{inc} for a certain material–environment combination is to use decreasing K (constant displacement) specimens and load them to various K_{I_i} values. By starting at high K_{I_i} levels the long incubation times can be avoided and yet the locus of t_{inc} can be found, as shown schematically in figure 10.9. Note that, as in figure 10.8, the ordinate is K_I, not K_{I_i}.

Crack growth rates, da/dt, depend on the material, environment and stress intensity factor K_I. Thus the crack growth period will depend strongly on

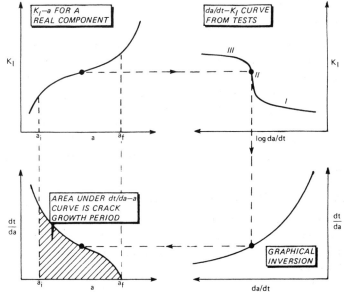

Figure 10.10. Schematic of integrating crack growth rate data from tests in order to predict the crack growth period for a real component.

specimen (or component) geometry and size, since these parameters control the variation of K with crack length. For example, if large and small specimens which are geometrically similar are loaded to the same K_{I_i}, the larger specimen takes longer to fail. In turn this means that an estimate of the crack growth period for a structural component must be obtained by integrating crack growth rate data. Figure 10.10 shows schematically how such an integration could be done.

Finally, the time to failure, t_f, for the component is obtained – at least in theory – by summing t_{inc} and the crack growth period. However, in practice there may be great difficulties. For example, the component geometry may be complex and stress intensity factors difficult to obtain. Also, stresses may relax as a crack grows, or else may redistribute owing to permanent changes in displacement (e.g. slip of bolted or riveted joints). Another important factor is the service environment, which may differ considerably from test environments.

10.6. Practical Significance of Sustained Load Fracture Testing

Threshold Stress Intensity, $K_{I_{th}}$

Most sustained load fracture tests are done with the primary aim of determining $K_{I_{th}}$. Although useful for specific applications, actual values of $K_{I_{th}}$ are not used as general design criteria because their significance for service performance has not been established. This is not surprising in view of the fact that $K_{I_{th}}$ depends on the material, environment, temperature, stress state (plane strain or plane stress) and often on very long testing times: the longer the testing times, the lower the apparent value of $K_{I_{th}}$.

Despite these problems $K_{I_{th}}$ is useful as a guide to material selection and design. For example, figure 10.11 shows stress corrosion $K_{I_{scc}}$ data for a

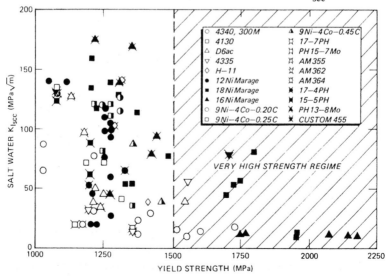

Figure 10.11. $K_{I_{scc}}$ for high strength steels, reference 5.

number of commercial high strength steels. In the very high strength regime $K_{I_{scc}}$ is low, irrespective of alloy type, composition, microstructure or heat treatment. This trend, in addition to the fact that stress corrosion crack growth rates are high, indicates that stress corrosion cracking cannot be tolerated by components made from very high strength steels. A fracture mechanics approach to design cannot be used.

Instead, the components must be rigorously inspected or proof loaded to ensure that they are free from macroscopic flaws, and thorough protection against corrosion must be applied (this is good practice, whatever the material).

(Proof loading is the application of a high load in the knowledge or expectation that if the component does not fail, it will then survive in service a given time, after which the proof load may be repeated to establish a further safe period of operation. There is always the possibility that the component will fail during proof loading. Understandably, therefore, this method of component verification is not very popular.)

As a development of the foregoing, an important contribution to interpretation of $K_{I_{th}}$ testing has been made by the U.S. Naval Research Laboratory, reference 6 of the bibliography. This is the concept of Ratio Analysis Diagrams (RADs) in which a grid of lines of constant K_I/σ_{ys} are superimposed on K_{I_c} and $K_{I_{th}}$ data plotted against σ_{ys}. An example of such a diagram for practical interpretation of K_{I_c} has already been given in section 5.6.

Figure 10.12 shows a K_{I_c} and $K_{I_{scc}}$ RAD for a high strength stainless steel. On this diagram the data zones for fast fracture (K_{I_c}) and initiation of stress corrosion crack growth ($K_{I_{scc}}$) are plotted with a grid of K_I/σ_{ys} lines that separate the diagram into regions of high, low and intermediate ratios according to the following rationale:

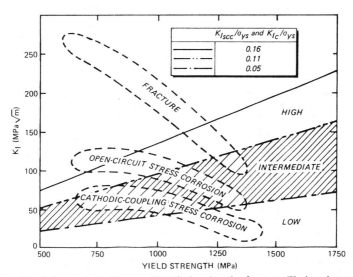

Figure 10.12. Ratio Analysis Diagram (RAD) showing the fracture (K_{I_c}) and stress corrosion ($K_{I_{scc}}$) properties of a high strength stainless steel, reference 6.

(1) The high ratio region is where high stresses and large cracks are necessary to cause fast fracture or stress corrosion crack growth. A fracture mechanics approach to design is relatively straightforward.

(2) The low ratio region is where fast fracture or stress corrosion crack growth can initiate from very small defects at moderate to low stress levels. This region is where a fracture mechanics approach to design cannot be used.

(3) The intermediate region is where the combination of high stresses and small flaws, low stresses and large flaws, or intermediate stress levels and flaw sizes are critical. This region is where a highly refined application of fracture mechanics is required for adequate design.

Crack Growth Rate Tests

Crack growth rate data can be useful in several ways:

- estimating $K_{I_{th}}$ from decreasing K tests, section 10.3
- predicting the service lives of components, provided that the incubation time t_{inc}, non-steady state crack growth, and the practical difficulties mentioned at the end of section 10.5 are absent or can be accounted for
- deciding whether a period of safe crack growth exists, and if so, determining inspection intervals for parts assumed or known to contain flaws.

For some material-environment combinations the crack growth rates measured in tests are so high that even though $K_{I_{max}}$ may be much higher than $K_{I_{th}}$, it is clear that there is no possibility of setting reasonable inspection intervals for a period of safe crack growth. Examples are liquid metal embrittlement of various materials and aqueous stress corrosion of high strength steels and titanium alloys. In these cases either the material must have a

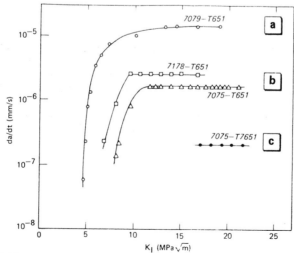

Figure 10.13. Outdoor exposure stress corrosion cracking propagation in 7000 series aluminium alloy plate, reference 7.

high $K_{I_{th}}$ value, or else sustained load fracture cannot be tolerated, as mentioned earlier.

On the other hand, important problems like stress corrosion cracking of high strength aluminium alloys can be better understood by analysis of crack growth rate data. Figure 10.13 shows crack growth rate data for several 7000 series aluminium alloys exposed to an outdoors environment of intermittent rain and moderate humidity. The data cover a range of possibilities for region II crack growth:

a) Fast crack growth in 7079−T651.
b) Fairly slow crack growth in 7075−T651 and 7178−T651.
c) Very slow crack growth in 7075−T7651.

As an example, consider components made from these alloys and installed in structures with feasible inspection intervals of a few months. Only in case (b) is the region II crack growth rate useful for estimating an inspection period of this time scale. In case (a) the region II crack growth rate is too high. In case (c) crack growth is so slow that inspection for cracks every few months is a waste of time.

10.7. Bibliography

1. Brown, B.F., *A New Stress-Corrosion Cracking Test for High Strength Alloys*, Materials Research and Standards, Vol. 6, pp. 129−133 (1966).
2. Chell, G.G. and Harrison, R.P., *Stress Intensity Factors for Cracks in Some Fracture Mechanics Test Specimens under Displacement Control*, Engineering Fracture Mechanics, Vol. 7, pp. 193−203 (1975)
3. Dorward, R.C. and Hasse, K.R., *Flaw Growth of 7075, 7475, 7050 and 7049 Aluminium Plate in Stress Corrosion Environments*, Final Report Contract NAS8−30890, Kaiser Aluminium and Chemical Corporation Centre for Technology (1976): Pleasanton California.
4. Wei, R.P., Novak, S.R. and Williams, D.P., *Some Important Considerations in the Development of Stress Corrosion Cracking Test Methods*, Materials Research and Standards, Vol. 12, pp. 25−30 (1972).
5. Wanhill, R.J.H., *Microstructural Influences on Fatigue and Fracture Resistance in High Strength Structural Materials*, Engineering Fracture Mechanics, Vol. 10, pp. 337−357 (1978).
6. Judy, R.W., Jr. and Goode, R.J., *Prevention and Control of Subcritical Crack Growth in High-Strength Metals*, Report 7780, Naval Research Laboratory (1970): Washington, D.C.
7. Hyatt, M.V. and Speidel, M.O., *Stress Corrosion Cracking of High-Strength Aluminium Alloys*, Report D6−24840, Boeing Commercial Airplane Group (1970): Seattle, Washington.

11. DYNAMIC CRACK GROWTH AND ARREST

11.1. Introduction

In this concluding chapter on fracture mechanics and crack growth the behaviour of statically loaded cracks growing beyond instability will be discussed.

The onset of instability usually means failure of a component or structure. As such, either instability cannot be allowed to occur, or else a partial (component) failure can be tolerated until its detection by regular inspection. In the latter case the component must be repaired or replaced if possible. Otherwise the structure has to be retired from service.

Certain structures belong to a category for which instability is always a possibility and failure would be intolerable. Examples are pipelines, nuclear reactor pressure vessels and liquid natural gas containers. For this category of structures the design and construction must include features to ensure crack arrest.

The foregoing examples illustrate the importance of unstable, i.e. dynamic, crack growth and its arrest. However, the study of this phenomenon is a highly specialised one, and is not amenable to detailed treatment in a basic course on fracture mechanics. In what follows only general remarks about the different concepts will be made.

In section 11.2 two basic aspects of dynamic crack growth will be described. First the velocities of fast fractures and the fact that they are finite. Second, as a consequence of finite crack velocity it is possible for crack branching to occur. Then in section 11.3 the conditions for crack arrest will be given together with their practical significance.

Section 11.4 is an overview of the widely used fracture mechanics methods of analysing dynamic crack growth and arrest in pipelines and thick-walled pressure vessels, and includes the concept of dynamic fracture toughness.

Finally, in sections 11.5 and 11.6 some information will be given about experimental determination of dynamic fracture toughness and the instantaneous value of the dynamic stress intensity factor.

11.2. Basic Aspects of Dynamic Crack Growth

There are two basic aspects of dynamic crack growth:
(1) Finite velocities of crack propagation.
(2) Crack branching: this falls in the category of macrobranching rather than microbranching (which commonly occurs during all kinds of crack growth).

Finite Velocities of Crack Propagation

Dynamic crack growth may be considered in terms of an energy balance. This will be shown with the help of figure 11.1, which is a simplified crack resistance $(G, R - a)$ diagram. After initiation of unstable crack extension there is excess energy which increases during crack growth. By the time the crack has reached a length a_i the total excess energy has amounted approximately to the shaded area in figure 11.1. (In practice G does not have to increase linearly with increasing crack length. Nor is it necessarily valid that R remains constant during dynamic crack growth.)

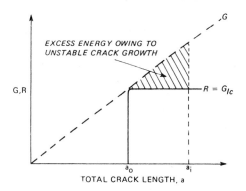

Figure 11.1. $G, R - a$ diagram showing the excess in energy some time after initiation of unstable crack extension under plane strain conditions.

The approximation in figure 11.1 is, however, convenient and adequate for further analysis to indicate that crack velocities are finite. The analysis which follows was first published by Mott, reference 1 of the bibliography to this chapter, as long ago as 1948.

Assuming that the formulation for G is the same at the onset of crack extension and during unstable fracture, the excess energy in figure 11.1 can be expressed by

$$U_a = \int_{a_0}^{a_i} (G - R)\, da = -R(a_i - a_0) + \int_{a_0}^{a_i} \frac{\pi \sigma^2 a}{E'}\, da. \qquad (11.1)$$

Note that since $R = G_{Ic}$ is assumed to be constant there is a condition of plane strain. Thus $E' = E/(1 - \nu^2)$. Also, R is given by $R = \pi \sigma^2 a_0/E'$. Substituting in equation (11.1) we obtain

$$U_a = -\frac{\pi \sigma^2 a_0}{E'}(a_i - a_0) + \frac{\pi \sigma^2}{2E'}(a_i^2 - a_0^2)$$

$$= \frac{\pi \sigma^2}{2E'}(a_i - a_0)^2. \qquad (11.2)$$

Mott argued that for a propagating crack the excess energy is stored as kinetic energy. Some 25 years later this postulate was experimentally confirmed by Hahn et al., reference 2.

A simple expression for the stored kinetic energy is obtainable from the opening displacement of the crack flanks. From section 2.3 the displacement v in the y direction is

$$v = \frac{2\sigma}{E'}\sqrt{a^2 - x^2}. \tag{11.3}$$

Since x is a function of a we may write $x = Ca$ for $0 < C < 1$. Then

$$v = \frac{2\sigma}{E'}\sqrt{a^2(1 - C^2)} = C_1\frac{\sigma_a}{E'}. \tag{11.4}$$

As the crack propagates the displacement v will change with time. Denoting the rate of change d/dt with "˙", we may write

$$\dot{v} = \frac{C_1 \sigma \dot{a}}{E'}. \tag{11.5}$$

The material adjacent to the crack flanks is displaced with a velocity $\dot{v}$. The kinetic energy in the displaced material is $T = mV^2/2$. For a material of density ρ and per unit thickness

$$T = \tfrac{1}{2}\rho \cdot area \cdot V^2 = \tfrac{1}{2}\rho \iint(\dot{v}^2)dxdy$$

$$= \tfrac{1}{2}\rho \dot{a}^2\frac{\sigma^2}{E'^2}\iint(C_1^2)dxdy. \tag{11.6}$$

The solution of the integral in equation (11.6) will have the dimension $[LENGTH]^2$. For a semi-infinite plate the only significant length is the crack length a. Thus the integral can be expressed as ka^2, and for a crack of length a_i (see figure 11.1)

$$T = \tfrac{1}{2}\rho\, \dot{a}^2 ka_i^2\frac{\sigma^2}{E'^2}. \tag{11.7}$$

If all the excess energy owing to unstable crack growth is converted into kinetic energy, then U_a in equation (11.2) will equal T. Thus

$$\frac{\pi\sigma^2}{2E'}(a_i - a_0)^2 = \frac{\rho k\sigma^2}{2E'}(\dot{a}a_i)^2$$

and

$$\dot{a}^2 = \frac{\pi}{k}\frac{E'}{\rho}(\frac{a_i - a_0}{a_i})^2$$

$$\dot{a} = \sqrt{\frac{\pi}{k}}\sqrt{\frac{E'}{\rho}}(1 - \frac{a_0}{a_i}) \tag{11.8}$$

The quantity $\sqrt{E'/\rho}$ is the velocity, V_g, of a longitudinal wave in a material. Consequently equation (11.8) can be rewritten as

$$\frac{\dot{a}}{V_g} = \sqrt{\frac{\pi}{k}}(1 - \frac{a_0}{a_i}). \tag{11.9}$$

For long propagating cracks $a_i \gg a_0$ and equation (11.9) has a limit value of $\sqrt{\pi/k}$. The limit value has been calculated to be much less than unity, and experimental measurements have provided confirmation of this. Thus it can be stated that the maximum velocity of a propagating crack will always be a fraction of the longitudinal wave speed.

More elaborate analyses have shown that for a brittle material the theoretical maximum crack velocity is equal to the velocity of a surface wave.

Crack Branching

Dynamic crack growth may be accompanied by multiple branching of the crack. Some authors have attempted to explain branching in terms of kinetic energy. But experiments using high speed cameras have shown that branching does not alter the crack velocity. This calls into question any analysis based on kinetic energy, since if kinetic energy were used for crack branching the velocity would decrease.

Such experiments have shown, however, that crack branching under mode I loading occurs only when a specific stress intensity factor is exceeded. An illustration of this is given in figure 11.2 for six glass plates containing crack starters in the form of notches of increasing sharpness in the order a – f. The blunter the notch the higher the stress required to initiate fracture. This means that at any crack length the stress intensity factor for the propagating crack is highest in specimen (a) and lowest in specimen (f). The result, as figure 11.2 shows, is that crack branching is greatest in specimen (a) and least in specimen (f).

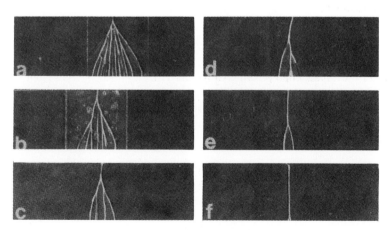

Figure 11.2. Crack branching in glass plates. *Courtesy J.E. Field, Cambridge University.*

The dependence of crack branching on the stress intensity factor can be explained qualitatively. When a specific stress intensity factor is exceeded new cracks initiate ahead of the main crack. At first such cracks initiate so close to the main crack that they are overtaken by it. Eventually the initiation of new cracks is sufficiently far ahead of the main crack that they can accelerate to a velocity whereby they are no longer overtaken. Branching then occurs.

Evidence for this explanation has been provided by experiments on brittle materials. In particular, the observation of roughening of the surfaces of the main crack shortly before branching is evidence that new cracks initiate ahead of the main crack and are at first overtaken by it.

Crack branching can be promoted by the finite geometry of specimens. This effect has its origin in the maximum crack velocity being less than the velocities of waves in the material. Shock waves caused by impact loading and/or dynamic crack growth travel faster than the crack. The shock waves reflect from the back surface of a specimen. On arriving back at the crack tip the shock waves suddenly increase the stress intensity factor. This often causes multiple branching: excellent examples are sometimes provided by broken window panes.

11.3. Basic Principles of Crack Arrest

Like dynamic crack growth, crack arrest can be considered in terms of an energy balance. In the first instance one might consider the problem of crack arrest as an energy rate balance, such that the crack stops if G somehow decreases below R. If R is constant this criterion is exactly the reverse of that for unstable crack extension (cf. section 4.5). The situation for plane strain conditions is depicted schematically in figure 11.3a.

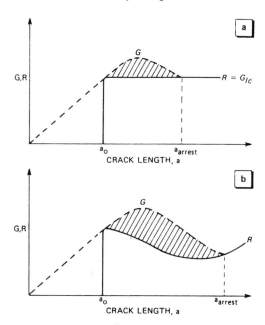

Figure 11.3. Crack arrest in terms of an energy rate balance (G = R at arrest) for plane strain conditions with R (a) insensitive and (b) sensitive to crack velocity.

However, for some materials it is very probable that R is not constant but depends on the crack velocity. The reason is as follows. The material in front of a fast running crack will be loaded at very high strain rates, and for strain rate sensitive materials the yield strength increases with increasing strain rate. In turn, a higher yield strength decreases the amount of crack tip plasticity and hence R, which is mainly plastic energy. In this case the energy rate balance criterion for crack arrest is as depicted in figure 11.3b.

The energy rate balance criterion is in fact an oversimplification. In section 11.2 it was argued that excess energy after instability is converted to kinetic energy. If this kinetic energy can be used for crack propagation the situations given in figure 11.3 are no longer valid, and the problem must be solved by equating the total amounts of energy and not just energy rates. This is shown schematically in figure 11.4 for plane strain conditions. Note that for a strain rate sensitive material (situation in figure 11.4b) R will increase as the arrest point is approached. This is because the decrease in kinetic energy will be accompanied by a decrease in crack velocity and hence lower strain rates and a lower yield strength ahead of the crack.

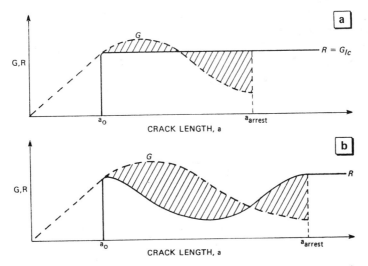

Figure 11.4. Crack arrest in terms of a balance of total energies for plane strain conditions with (a) R insensitive and (b) R sensitive to crack velocity.

Although it is inconceivable that all the kinetic energy of a fast running crack is available for crack propagation, the situations shown in figure 11.4 are generally considered representative of actual crack arrest behaviour in many materials, since they are qualitatively consistent with experimental results.

A potentially important consequence of crack arrest depending (approximately) on a balance of total energies is that the value of G at arrest is not a material constant, since it will depend on the variation of both G and R with crack length and velocity. This is exemplified in figure 11.5, which shows that even for the same maximum value of G and a constant R, the value of G at arrest can be clearly different for different initial (and hence final) crack lengths. This problem will be mentioned again in section 11.4 in terms of the static stress intensity factor at crack arrest.

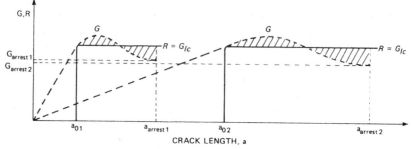

Figure 11.5. Effect of initial crack length on G_{arrest} for a strain rate insensitive material under plane strain conditions.

The Practical Significance of Crack Arrest

The foregoing considerations, even though qualitative, are sufficient to demonstrate that crack arrest in a structure can occur if the energy release rate, G, decreases and/or the crack resistance, R, increases.

A decrease in G is obtained if

- the crack grows into a decreasing stress field, as in the case of wedge loading (see section 5.4)
- the load causing instability is transient and decreases with time
- part of the load on the cracked element is taken up and transmitted by other structural elements: this is also referred to as load shedding by the cracked element.

A simple example of the load being taken up and transmitted by another structural element is given in figure 11.6. This shows a cracked plate with a

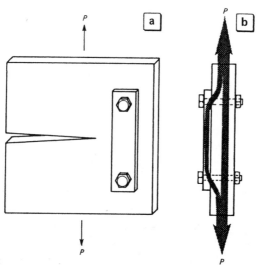

Figure 11.6. Cracked plate with bolted on arrest strip.

bolted on arrest strip. As the crack approaches, the arrest strip increasingly resists the normal displacement of the crack flanks. Consequently, part of the load on the plate is taken up and transmitted by the arrest strip, as indicated schematically in figure 11.6b. This principle is much used (in a more sophisticated way) in aircraft structures, and serves the dual purpose of arresting fast running cracks and slowing down fatigue crack growth. More information on this subject is given in Broek's book on fracture mechanics, reference 3 of the bibliography to this chapter.

Another crack arrest configuration, this time for a pipeline, is shown in figure 11.7. The arrest of fast running cracks in pipelines is extremely important. It has been known for fast fractures in gas pipelines to run for kilometres, with disastrous consequences. Pipeline fracture is further discussed in section 11.4.

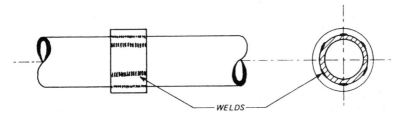

Figure 11.7. Pipeline with welded on crack arrest ring.

An increase in a material's crack resistance, R, is not easily obtained. The only possibility that has practical significance is to dimension the structure such that instability would be accompanied by a change from plane strain to plane stress conditions. This would result in a rapidly rising R-curve (see section 4.6) which could soon cause crack arrest even if G continued to increase. This is shown schematically in figure 11.8.

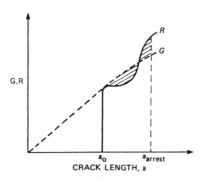

Figure 11.8. Increase in crack resistance owing to a transition from plane strain to plane stress.

There are other possibilities of increasing R. They have very limited applicability. Examples are direct insertion of strips of laminated material or strips of a material with much higher fracture toughness (and usually a much lower yield strength). These possibilities are feasible for weldable materials, but apart from technical difficulties there is a major manufacturing problem, since to be truly effective the strips must be spaced fairly close together.

11.4. Fracture Mechanics Analysis of Fast Fracture and Crack Arrest

Fracture mechanics analysis of fast fracture and crack arrest is a highly specialised topic, and a comprehensive treatment of the subject is beyond the scope of this course. Instead we shall make some general remarks concerning the two main problem areas, pipelines and thick-walled pressure vessels, for which dynamic fracture and crack arrest must be considered.

Pipelines

As mentioned in the previous section, the arrest of fast running cracks in pipelines is extremely important, since fast fractures have been known to run for kilometres. The problem is especially severe for gas pipelines, for the following reason. When a gas pipeline fractures the gas depressurises rapidly to cause a decompression shock wave travelling at the speed of sound (≈ 400 m/s) in the gas. If the fast fracture in the pipe is able to travel faster than the decompression shock wave (as is quite possible) the crack tip continues to run in fully loaded material and there is no chance, without crack arrest rings, for crack arrest to occur.

On the other hand, the pressure in a liquid-filled pipeline will drop immediately if the pipeline breaks open, since liquids are virtually incompressible. The drop in pressure will cause an immediate decrease in the load acting on the pipe wall and consequently a decrease in G and the stress intensity of the crack. This decrease may well be sufficient for a running crack to arrest.

Early studies of the fast fracture problem in pipelines were based on the assumption that the crack arrest will occur at a characteristic stress intensity (see reference 4 for more detailed information). However, the problem is currently viewed in terms of a dynamic energy balance. In general terms, crack propagation continues as long as

$$G_{dyn}(a,t) \geqslant R(\dot{a}) \qquad (11.10)$$

where t denotes time; and for unit thickness

$$G_{dyn} = \frac{1}{B}(\frac{dF}{da} - \frac{dU_a}{da} - \frac{dT}{da}). \qquad (11.11)$$

Equation (11.11) differs from the expression for the static energy release rate, G, cf. equation (4.4), by the addition of a kinetic energy term, dT/da.

Practical application of a dynamic energy balance criterion to fast fracture of a gas pipeline requires the following aspects to be considered:
- the work done by the pressurized gas on the pipe walls as they crack
- the contribution of kinetic energy to crack growth
- the inertia of the pipe walls that have already cracked

- the decrease of gas pressure owing to leakage
- the effect of the large amount of plastic deformation of the pipe walls behind the crack tip (the pipe walls bend outwards, see figure 11.9)
- possible constraint of the plastic deformation of the pipe walls owing to the pipe being covered by soil (generally called "backfill").

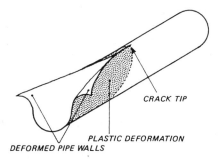

Figure 11.9. Schematic of fast fracture in a pipe.

In reference 6 a simplified (but still complex) expression for the dynamic strain energy release rate was obtained with the assumption of "steady state" crack propagation, i.e. G_{dyn} independent of crack length and dependent only on time and hence crack velocity. The variation of G_{dyn} with crack velocity was calculated using equation (11.11). An example is given in figure 11.10.a, showing that G_{dyn} strongly depends on crack velocity. For this particular example G_{dyn} reaches a maximum value $\sim$6.4 MJ/m^2 when there is no soil coverage. Thus if the pipe material has a minimum crack resistance, R_{min}, of 6.5 MJ/m^2 it is not possible for fast fracture to achieve a steady state of continuing crack propagation: i.e. if fast fracture somehow initiates it will arrest, even when crack arrest rings are not used.

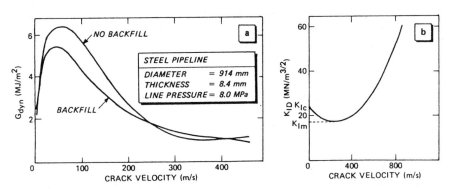

Figure 11.10. a) G_{dyn} calculated as a function of crack velocity. After reference 6
b) K_{ID} measured as a function of crack velocity for a pressure vessel steel.

If it is at all possible to calculate G_{dyn} for a fast fracture situation, then it is evidently important to obtain data for the dynamic crack resistance, $R(\dot{a})$, or the equivalent dynamic fracture toughness, $K_{ID}(\dot{a})$. The criterion for crack arrest being inevitable is that G_{dyn} remains less than R_{min} or that K_I^{dyn} becomes less than K_{IDmin}, where the minimum in the variation of K_{ID} with crack velocity is usually denoted as K_{Im}. Figure 11.10b shows that K_{ID} and hence R can also strongly depend on crack velocity. Note that as in the static case (section 4.8) we can express the dynamic crack resistance curve in terms of stress intensity factors. Estimates of K_{ID} are possible by assuming that during steady state crack propagation $G_{dyn} = R(\overset{\bullet}{a})$ and that

$$(\frac{K_{ID}}{K_{Ic}})^2 = (\frac{R(\overset{\bullet}{a})}{G_{Ic}}) = \frac{G_{dyn}}{G_{Ic}} \tag{11.12}$$

where G_{dyn} follows from equation (11.11), the solution of which, however, requires a complete dynamic analysis of the fracture problem. Alternatively, K_{ID} can be determined experimentally. This is discussed in section 11.5.

In fact, for steel pipelines the problem is generally solved in a more empirical way. The energy release rate, G_{dyn}, is correlated with the energy, C_V, of conventional Charpy Impact tests carried out above the brittle-to-ductile transition temperature (figure 1.2). The minimum crack resistance required for a pipeline steel is then expressed as a minimum required Charpy energy, C_{Vmin}.

Correlations between G_{dyn} and C_V have no physical background, but they are used because the Charpy Impact test is still the most common method of determining a steel's resistance to brittle fracture. The correlations are carried a step further by establishing empirical relations between C_{Vmin} and the main parameters, besides the crack velocity, that determine the maximum value of G_{dyn}. These parameters are the pipe radius, R, wall thickness, B and line pressure P.

Several empirical relations for C_{Vmin} exist. Examples are

(ref. 6) $\quad C_{Vmin} = 3.36 \cdot 10^{-4}(\sigma_H^{1.5} \cdot R^{0.5})$ $\qquad$ (11.13)

(ref. 7) $\quad C_{Vmin} = (1.713\frac{R}{B^{0.5}} - 0.2753\frac{R^{1.25}}{B^{0.75}})\sigma_H \cdot 10^{-3}.$ $\qquad$ (11.14)

In these equations C_V is expressed in Joules; σ_H in Pa $(= N/m^2)$; and R and B are in mm. σ_H is the hoop stress $(= PR/B)$ in the pipe. Such empirical relations seem to work reasonably well. Nevertheless, there has been much recent investigation of actual dynamic fracture toughness with a view to improving the understanding and hence prediction of crack arrest.

Thick-Walled Pressure Vessels

Besides pipelines, dynamic fracture and crack arrest are also of concern for thick-walled (nuclear) pressure vessels, but with major differences. The critical cracks to be considered are part through flaws growing from the inside surface of the pressure vessel wall. An example has been given at the end of section 2.8 in chapter 2. Cracks may become unstable owing to thermal shock (unusually rapid cooling within the pressure vessel) whereby high tensile stresses are induced at the inside surface of the wall.

It is essential that the cracks stop soon after instability so that wall penetration does not occur. The main reason why crack arrest is possible is the steep negative stress gradient through the wall to the outside surface. An increase in fracture toughness due to a lesser amount of radiation damage can be an additional factor in a nuclear pressure vessel.

Since there is only a very limited amount of permissible dynamic crack growth, the cracks will not have much kinetic energy. This fact is reflected in one of the methods developed to analyse dynamic fracture in pressure vessels steels, the "crack arrest toughness" approach of Crosley and Ripling (reference 8).

Crosley and Ripling assert that when the amount of unstable crack growth is very limited the dynamic effects on stress intensity can be neglected. If this is the case, then the calculated static stress intensity factor, K_{Ia}^{stat} for a crack that has just arrested should be a reasonable approximation to the actual stress intensity at crack arrest. For this concept to be useful it has to be demonstrated that K_{Ia}^{stat} is approximately constant within the appropriate range of crack lengths and velocities, and that there is in fact little difference between K_{Ia}^{stat} and the value of K_I^{dyn} at crack arrest, K_{Ia}^{dyn}.

For several years there has been considerable controversy over the appropriateness of the static approach advocated by Crosley and Ripling. In particular, Kanninen and coworkers (e.g. reference 2) suggest that a dynamic energy balance approach should be used. The controversy has led to much effort in determining dynamic fracture toughness in order to compare the static and dynamic approaches to crack arrest.

Kalthoff et al. (reference 9) have made a summary comparing the results of the static and dynamic approaches in a three-dimensional graph, shown in slightly modified form in figure 11.11. The K_I versus crack length curves are

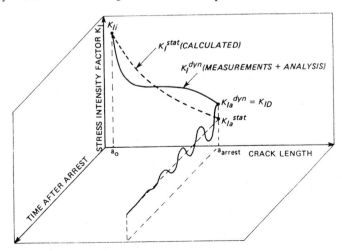

Figure 11.11. Comparison of the static and dynamic approaches to crack arrest in a double cantilever beam specimen (DCB).

schematic for the situation of crack arrest in a double cantilever beam specimen (DCB), but this does not detract from the generality of comparison.
During instability K_I^{dyn} is at first lower than K_I^{stat}. However, as the arrest
point is approached K_I^{dyn} becomes greater than K_I^{stat}. At arrest K_{Ia}^{dyn} is equal
to the dynamic fracture toughness, K_{ID}, which is greater than K_{Ia}^{stat}.
Note that K_I^{dyn} oscillates with time after arrest and eventually becomes K_{Ia}^{stat}.
At arrest K_{Ia}^{stat} is a good approximation of K_{Ia}^{dyn} only if the oscillations are
limited and damp out quickly.

The difference between K_{Ia}^{dyn} and K_{Ia}^{stat} increases with increasing amount
of unstable crack growth, $a_{arrest} - a_0$. Also, this difference depends on the
specimen geometry: for DCB specimens the difference is significant, but for
crack line wedge-loaded specimens (CLWL) the difference is usually small.

At present it is still not clear whether K_{Ia}^{stat} is adequate for predicting crack
arrest in thick-walled pressure vessels, although there does appear to be a preference for this approach. The main reasons for this preference are:

(1) Many investigations support the K_{Ia}^{stat} approach when the amount of
unstable crack growth is very limited, as it has to be in pressure vessels.
(Thus the potentially important variation in G_{arrest}, and hence K_{Ia}^{stat},
illustrated in figure 11.5 is not significant for unstable crack growth
and arrest in pressure vessels.)

(2) Results of a cooperative test programme carried out by ASTM members
between 1977 and 1980 showed that the data could be described fairly
well by a constant K_{Ia}^{stat} value, figure 11.12, despite wide variation in
the initial stress intensity, K_{Ii}, causing instability. As illustrated in section 5.4, a higher K_{Ii} means a higher initial crack driving force. This
results in a higher crack velocity and more kinetic energy, so that if
dynamic effects had been significant the constancy of K_{Ia}^{stat} would not
have been observed.

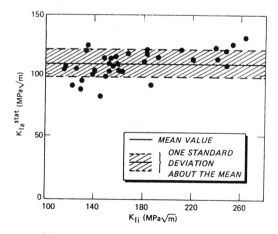

Figure 11.12. K_{Ia}^{stat} versus K_{Ii} for SA533 pressure vessel steel. After reference 10.

(3) Given a reasonable degree of succes for the static approach, it is obviously much simpler to use a static value of the stress intensity factor than to carry out a dynamic analysis.

11.5. Determination of Dynamic Fracture Toughness, K_{ID}

As yet there is no standard test procedure for dynamic fracture toughness testing. But in view of the considerable progress in experimental and analytical techniques in the last few years, it can be safely assumed that a test standard will be available in the near future.

Several types of specimen have been used in the past, including double cantilever beams with and without taper (DCB and TDCB), single edge notched plates (SEN) and crack line wedge-loaded specimens (CLWL). Preference is now given to the CLWL arrangement shown in figure 11.13, for the following reasons:

- wedge loading, rather than pin loading, limits dynamic energy exchanges between specimen and testing machine
- the CLWL specimen is relatively economical of material: this is an important consideration for high toughness materials like pressure vessel steels, since large specimens are normally required (but see next paragraph for further details)
- the amount of side grooving to ensure an in-plane fracture path is much less than that for DCB specimens.

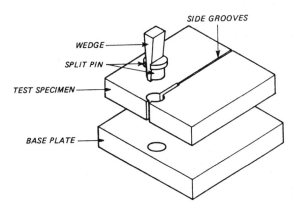

Figure 11.13. Schematic of the crack line wedge-loading arrangement for dynamic fracture toughness testing.

Even though the CLWL specimen is relatively economical of material, the requirement of K_{Ii} values well above K_{ID} would mean impossibly large specimens for high toughness materials. This problem has been avoided in two ways, illustrated in figure 11.14. In one case a CLWL specimen similar in shape to that shown in figure 5.17 has a brittle weld bead at the tip of the crack starter slot. The other specimen type is duplex, with crack initiation occurring in a hardened steel starter section welded onto the test material.

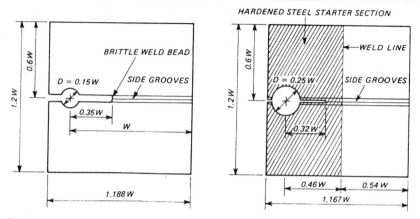

Figure 11.14. Crack line wedge-loaded specimens for dynamic fracture toughness testing.

Both types of specimen have a blunt starter notch, so that the stress intensity to initiate cracking will be higher than K_{ID}.

Test Procedures and Analysis

Only an outline of the test procedures and analysis will be given here, since there is as yet no standard procedure. There are two basic measurements to be made:

(1) A load-displacement record in order to determine K_{Ii}. Wedge-loading can be done at a relatively slow rate, so that recording the load signal from the testing machine and the displacement from a clip gauge mounted on the specimen is not a special problem. (A load displacement record after crack arrest can also be made. This serves as a check on the first measurement, but is not strictly necessary, since stiff wedge-loading results in fracture with almost constant displacement at the load line.)

(2) Measurement of crack length at arrest. Since the specimens are fairly thick (B ~ 50 mm) the through thickness crack length must be measured and averaged. This can be done by heat tinting. The cracked but unbroken specimens are heated in a furnace to discolour the fracture surfaces by oxidation. Subsequently the specimens are broken open and the crack lengths measured directly.

In addition, there are several techniques for measuring the crack length as a function of time during fast fracture in order to derive the crack velocity. These techniques are summarised in reference 11. They are all rather specialised and not suitable for use on a routine basis. One of these techniques, high speed photography, is used with other optical methods to determine dynamic stress intensity factors, discussed in section 11.6.

The values of K_{Ii} (determined from static analysis) and the amount of unstable crack growth ($a_{arrest} - a_0$) are sufficient to determine K_{ID} when a dynamic analysis of the specimen is available. This is because the analysis gives single valued relations between $a_{arrest} - a_0$ and K_{ID}/K_{Ii} as well as between $a_{arrest} - a_0$

and crack speed $\dot{a}$. The latter relation is generally presented in terms of $\dot{a}/C_0$ where C_0 is the longitudinal wave velocity.

The results of the dynamic analysis for the specimens shown in figure 11.14 are given in figure 11.15 (after reference 12). Generally the minimum value of K_{ID}, K_{Im}, is determined from a number of measurements at various crack speeds (compare figure 11.10b).

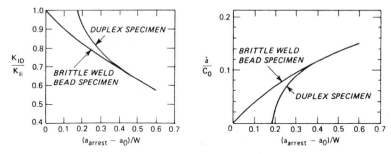

Figure 11.15. Reference curves for the assessment of crack speed $\dot{a}$ and dynamic fracture toughness K_{ID} in a CLWL specimen.

It is of course also possible to calculate K_{Ia}^{stat} from the load-displacement records, the total crack length at arrest, and the conventional static stress intensity factor calibration of the specimen. The values of K_{Ia}^{stat} and K_{ID} may then be compared to check for significant differences.

11.6. Determination of Dynamic Stress Intensity Factors

Calculation of dynamic stress intensity factors, K_I^{dyn}, is a very difficult problem that has had only limited success. Consequently, actual determination usually involves a combination of experimental measurements and analysis. Two such methods suitable for metallic materials are

- the shadow optical method of caustics in reflection, reference 9
- dynamic photoelasticity of birefringent coatings on specimens, reference 13.

Both methods require high speed photography to record instantaneous positions of fast running cracks.

The Shadow Optical Method of Caustics

The physical basis of this method is illustrated in figure 11.16 for the case of a transparent cracked specimen illuminated by parallel light. Owing to the stress concentration at the crack tip the specimen will locally contract and the refractive index will change. As a consequence the incident light will be deflected outwards during transmission. The overall effect is to produce a shadow spot bounded by a bright ring (the caustic) on an image plane at any distance behind the specimen.

A similar effect is obtainable from a nontransparent (e.g. metallic) specimen acting as a mirror. In this case the light being reflected forms an analo-

222

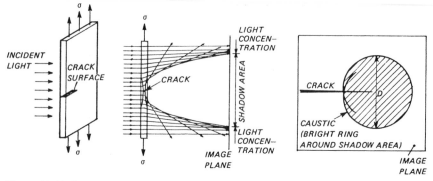

Figure 11.16. Schematic of the shadow optical method of caustics for transmitted light.

gous shadow spot and caustic when observed in a virtual image plane behind the specimen.

In both transmission and reflection arrangements the shadow spots for fast running cracks can be recorded by a high-speed camera focused on the appropriate image plane. The instantaneous dynamic stress intensity factor for a running crack is obtained from

$$K_I^{dyn} = F(\mathring{a})MD^{5/2} \qquad (11.15)$$

where D is the diameter of the caustic; $F(\mathring{a})$ is a factor accounting for the crack velocity dependence of the shadow spot and is $\sim$1 for typical fast fractures; and M accounts for specimen thickness, elastic and optical properties, and the distance between the specimen and the image plane. More details on the values of $F(\mathring{a})$ and M are given, for example, in reference 14.

Using this technique Kalthoff et al. (reference 9) have obtained K_I^{dyn} versus crack length curves for a number of materials, including a high strength steel. The results enabled them to indicate the general trend shown in figure 11.11.

Dynamic Photoelasticity of Coatings on Specimens

This method is an extension of the study of dynamic crack growth in transparent birefringent polymers. It has its basis in the fact that cracks and notches in birefringent polymers give rise to characteristic patterns of isochromatic fringe loops, whose size and shape can be analysed to derive the stress intensity factor when the loading conditions and optical properties of the polymers are known. This is a highly specialised topic: the interested reader is referred to e.g. reference 15 for information on such analyses.

To use the method for studying fast fracture in metallic specimens a birefringent coating is bonded on to a side surface as shown in figure 11.17 for a CLWL specimen. (Note that the coating is not continuous. This eliminates uncertainty as to whether the isochromatic pattern in the coating is influenced by fracture of the coating itself.) A dynamic fracture toughness test is then carried out in synchronization with high-speed photographic recording of the

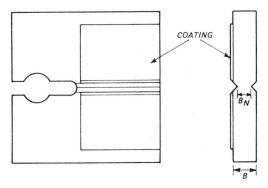

Figure 11.17. Schematic of birefringent coating on a crack line wedge-loaded specimen (CLWL). After reference 13.

instantaneous crack length in the specimen and the associated isochromatic fringe loops in the coating.

For any recorded instantaneous crack length in the metallic specimen the dynamic stress intensity factor is

$$(K_I^{dyn})_m = (\frac{E_m}{E_c})(\frac{1 + \nu_c}{1 + \nu_m})(K_I^{dyn})_c \sqrt{\frac{B}{B_N}} \qquad (11.16)$$

where the subscripts c and m refer to coating and metallic specimen respectively, and $(K_I^{dyn})_c$ for the coating has to be obtained by the rather complex analysis of the isochromatic fringe loops. The factor $\sqrt{B/B_N}$ is included in equation (11.16) because of the use of side grooves to ensure an in-plane fracture path in the specimen.

11.6. Bibliography

1. Mott, N.F., *Fracture of Metals: Theoretical Considerations*, Engineering, Vol. 165, pp. 16–18 (1948).
2. Hahn, G.T., Hoagland, R.G., Kanninen, M.F. and Rosenfield, A.R., *The Characterization of Fracture Arrest in a Structural Steel*, Pressure Vessel Technology, American Society of Mechanical Engineers, Part II, pp. 981–994 (1973): New York.
3. Broek, D., Elementary Engineering Fracture Mechanics, Martinus Nijhoff (1982): The Hague.
4. Maxey, W.A., Kiefner, J.F., Eiber, R.J. and Duffy, A.R., *Ductile Fracture Initiation, Propagation and Arrest in Cylindrical Vessels*, Fracture Toughness, ASTM STP 514, pp. 70–81 (1972): Philadelphia.
5. Kanninen, M.F., *An Analysis of Dynamic Crack Propagation and Arrest for a Material Having a Crack Speed Dependent Fracture Toughness*, Prospects of Fracture Mechanics, Noordhoff, pp. 252–260 (1974): Leyden.
6. *Running Shear Fracture in Line Pipe*, AISI Technical Report, American Iron and Steel Institute, September 1974.
7. Fearnehough, G.D., *Fracture Propagation Control in Gas Pipelines: a Survey of Relevant Studies*, International Journal of Pressure Vessels and Piping, Vol. 2, pp. 257–281 (1974).
8. Crosley, P.B. and Ripling, E.J., *Characteristics of a Run Arrest Segment of Crack Extension*, Fast Fracture and Crack Arrest, ASTM STP 627, pp. 203–227 (1977): Philadelphia.

9. Kalthoff, J.F., Beinert, J., Winkler, S. and Klemm, W., *Experimental Analysis of Dynamic Effects in Different Crack Arrest Test Specimens*, Crack Arrest Methodology and Applications, ASTM STP 711, pp. 109–127 (1980): Philadelphia.

10. Crosley, P.B. and Ripling, E.J., *Significance of Crack Arrest Toughness* (K_{Ia}) *Testing*, Crack Arrest Methodology and Applications, ASTM STP 711, pp. 321–337 (1980): Philadelphia.

11. Weiner, R.J. and Rogers, H.C., *A High-Speed Digital Technique for Precision Measurement of Crack Velocities*, Fast Fracture and Crack Arrest, ASTM STP 627, pp. 359–371 (1977): Philadelphia.

12. Hoagland, R.G., Rosenfield, A.R., Gehlen, P.C. and Hahn, G.T., *A Crack Arrest Measuring Procedure for* K_{Im}, K_{ID} *and* K_{Ia} *Properties*, Fast Fracture and Crack Arrest, ASTM STP 627, pp. 177–202 (1977): Philadelphia.

13. Kobayashi, T. and Dally, J.W., *Dynamic Photoelastic Determination of the* $\overset{.}{a}$ – K *Relation for 4340 Alloy Steel*, Crack Arrest Methodology and Applications, ASTM STP 711, pp. 189–210 (1980): Philadelphia.

14. Beinert, J., Kalthoff, J.F. and Maier, M., *Neuere Ergebnisse zur Anwendung des Schattenfleckverfahrens auf stehende und schnell-laufende Brüche*, 6th International Conference on Experimental Stress Analysis, VDI-Verlag GmbH, pp. 791–798 (1978): Düsseldorf.

15. Kobayashi, A.S., *Photoelasticity Techniques*, Experimental Techniques in Fracture Mechanics, Society for Experimental Stress Analysis, Iowa State Press, Chapter 6, Monograph No. 1, pp. 126–145 (1973): Ames, Iowa.

Part V Mechanisms of Fracture in Actual Materials

12. MECHANISMS OF FRACTURE IN METALLIC MATERIALS

12.1. Introduction

Since World War II there has been great progress in understanding the ways in which materials fracture. Such knowledge has proved essential to better formulation of fracture mechanisms. Nevertheless, it is still not possible to use this knowledge, together with other material properties, for predicting fracture behaviour in engineering terms with a high degree of confidence.

Some insight into the problems involved is given in chapter 13, and it is the intention of the present chapter to provide the necessary background information on fracture mechanisms.

Metallic materials, especially structural engineering alloys, are highly complex. An indication of this complexity is given by figure 12.1, which shows various microstructural features (not all of which need be present in a particular material) and also the two main types of fracture path, transgranular and intergranular fracture. Of fundamental importance is the fact that almost all structural materials are polycrystalline, i.e. they consist of aggregates of grains, each of which has a particular crystal orientation. The only exceptions are single crystal turbine blades for high performance jet engines.

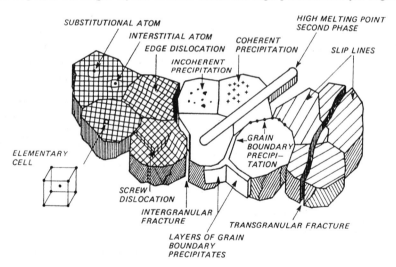

Figure 12.1. Schematic of microstructural features in metallic materials.
Courtesy Gerling Institut für Schadenforschung und Schadenverhütung, Cologne, FRG.

Before the various mechanisms of fracture are discussed some information will be given in sections 12.2 and 12.3 on the following topics:

(1) The instruments used in fractography, which is the study of fracture surfaces. In particular, the use of electron microscopes will be mentioned.

(2) The concept of dislocations (see figure 12.1). The nucleation and

movement of dislocations causes shear, i.e. slip, on certain sets of crystal planes, and the overall effect of slip is plastic deformation.

Sections 12.4–12.7 cover specific aspects of transgranular and intergranular fracture, namely

- transgranular fracture:
- ductile fracture by microvoid coalescence
- brittle fracture (cleavage)
- fatigue crack initiation and growth
- intergranular fracture:
- grain boundary separation with microvoid coalescence
- grain boundary separation without microvoid coalescence.

Finally, in section 12.8 some of the types of fracture that can occur owing to sustained loading at elevated temperatures (creep) or in aggressive environments (e.g. stress corrosion cracking) will be discussed.

12.2. The Study of Fracture Surfaces

Fracture surfaces exhibit both macroscopic and microscopic features, the study of which requires a wide range of magnification and a diversity of instruments, figure 12.2. Macroscopic examination should always be done first. This can be done with the unaided eye or a hand lens, and is often sufficient to indicate the directions in which cracks have grown and the locations of crack origins. Also, it is sometimes possible to distinguish immediately between fatigue and overload failures and whether the fracture is relatively recent. Older fracture surfaces tend to be discoloured owing to corrosion.

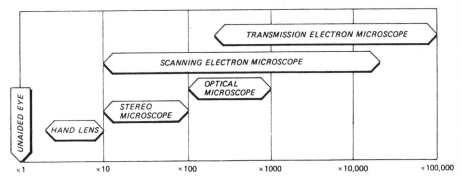

Figure 12.2. Approximate ranges of magnification for instruments used to study fracture surfaces and microstructure.

If the location of a crack origin is known, a stereo microscope is most useful for seeing whether there are special features associated with the origin. When this has been done the electron microscopes have to be used, especially the scanning electron microscope, which has a large depth of field and can be used from low to high magnifications. The scanning electron microscope has, in fact, virtually replaced the optical microscope for direct examination

228

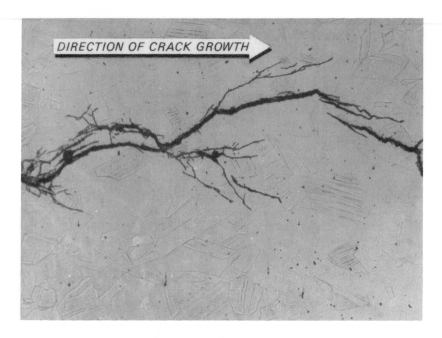

Figure 12.3. Transgranular stress corrosion crack in AISI 304 stainless steel. Optical metallograph, ×320.

of fracture surfaces. However, the optical microscope is indispensable for metallography, including polished and etched cross-sections through fracture surfaces and cracks. An example is given in figure 12.3.

The use of electron microscopy is essential for determining the types of fracture with certainty. This is because characteristic, identifying features are often revealed only at magnifications ~× 1000 or higher. Since the use of electron microscopes is somewhat specialised, the principles of their operation will be discussed briefly here.

The Transmission Electron Microscope (TEM)

The transmission electron miscroscope came into routine use in the 1950s. Broadly speaking, it can be compared to a photographic enlarger, using an electron beam and electromagnetic lenses instead of light and optical lenses.

Figure 12.4 gives a schematic of a TEM. Note that the electron beam passes through the specimen in the microscope (like light through the negative in an enlarger). Electrons can pass through a material thickness of only a few tens of nanometres (1 nm = 10^{-9} m). This means that the specimen must be very thin. For fractography it is therefore necessary to make a thin film-type replicate specimen of the actual fracture surface. A technique for preparing fracture surface replicas is illustrated in figure 12.5. The relatively thick copper grid is required to support the replica and enable its insertion in the TEM.

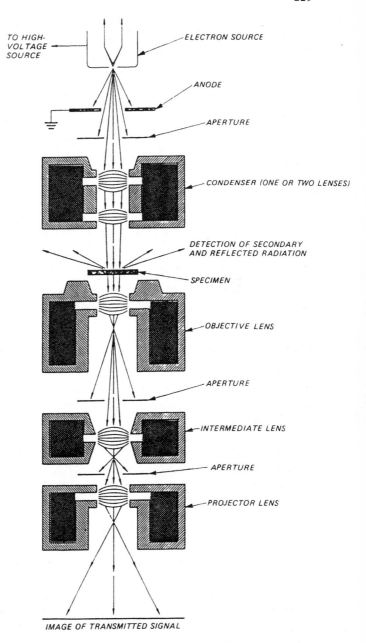

Figure 12.4. Schematic of a TEM.

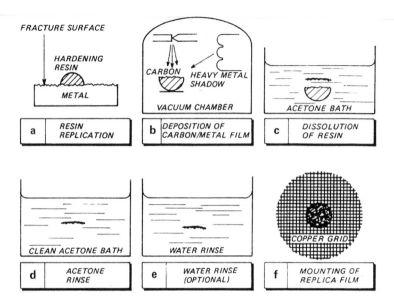

Figure 12.5. Schematic of preparation of a fracture surface replica.

The advantages of the TEM for fractography are that fracture surface details can be studied at very high magnifications (see figure 12.2) with good resolution and high contrast. Disadvantages include the time and skill required to prepare good replicas, the obscuring of about 50% of the replica by the supporting copper grid, and the fact that the replicas cannot be larger than a few millimetres in any direction.

The Scanning Electron Microscope (SEM)

The scanning electron microscope is a later development than the TEM and came into routine use in the late 1960s. The operating principle of a SEM is completely different from that of a TEM.

Figure 12.6 gives a schematic of a SEM. The electron beam is highly focussed by two condenser lenses and impacts the specimen as a small spot. The impact results in the generation of various forms of radiation, figure 12.7, of which the backscattered and secondary electrons are important for image forming.

The backscattered and secondary electrons come from different zones near the surface of the specimen, as shown in figure 12.8. The secondary electron emission zone is much smaller than the zone for backscattered electrons. Consequently, higher resolutions are obtained from images formed by secondary electron collection, amplification and display (see figure 12.6) and it is this secondary electron mode of operation that is normally used for fractography.

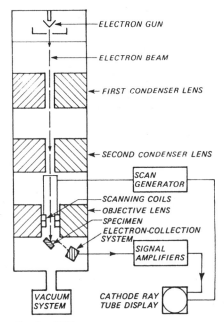

Figure 12.6. Schematic of a SEM.

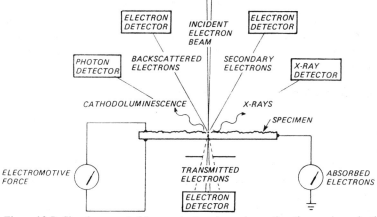

Figure 12.7. Signals generated by an electron beam impacting the specimen in the SEM.

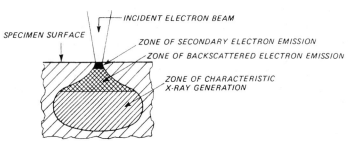

Figure 12.8. Zones in a specimen that are sources for signals generated by an electron beam impacting the specimen.

Actual operation requires the electron beam to scan rapidly back and forth in a systematic manner over the specimen surface. This is achieved using scanning coils in an objective lens, see figure 12.6. The scanning coils are controlled by a raster scan generator, which simultaneously controls the deflection coils of the cathode ray tube display. In this way the signal from the electron collector is displayed in the same pattern as it was generated by the electron beam impacting the specimen.

The relative intensities of the generated secondary electrons form an image of the specimen. Magnification in the SEM is changed by keeping the display size constant and changing the area of specimen scanned. Typical magnifications for fractographic studies range from × 10 to × 20,000, see figure 12.2.

The main advantages of using a SEM for fractography are that the specimen is observed directly and the image has a realistic three-dimensional appearance. Also, modern microscopes can accommodate fairly large specimens with maximum dimensions ~100 mm. These advantages have made the SEM the preferred instrument for fractography, despite the better resolution of the TEM. However, both instruments should be used for fracture surface examination, especially when service failures are involved, in order to maximise the chance of identifying characteristic features of different types of fracture.

12.3. Slip, Plastic Deformation and Dislocations

The most common form of plastic deformation is slip. This is the shearing of whole blocks of crystal over one another, see figure 12.9. Thus plastic deformation is intrinsically inhomogeneous. Slip always occurs by displacement of blocks of crystal in specific crystallographic directions on particular sets of crystal planes called slip planes. When slip occurs on several sets of crystal planes in each grain of a polycrystalline material the overall result is plastic deformation that is effectively homogeneous. This is why fracture mechanics concepts, which are based on continuum behaviour, can be used to describe many aspects of crack extension and fracture in structural materials.

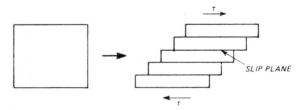

Figure 12.9. Schematic of plastic deformation (shear) by slip.

The displacement of crystal blocks does not occur simultaneously over the entire slip plane. Instead it occurs consecutively, beginning in a very small region on the slip plane and spreading outwards. The boundary between the regions where slip has and where slip has not occurred is called a dislocation and is commonly represented as a line in the slip plane. The two basic types of dislocation and slip displacements associated with them are shown in figure 12.10. When the displacement is perpendicular to the dis-

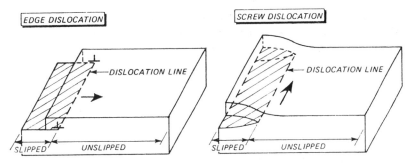

Figure 12.10. Slip owing to movement of pure edge and pure screw dislocations.

location line it is called an edge dislocation. If the displacement is parallel to the dislocation line it is of the screw type. In practice, most dislocation lines are neither pure edge nor pure screw but are mixtures of the two.

Note the symbol ⊥ used for an edge dislocation. This symbol and its inverse represent the fact that an edge dislocation can be depicted as an extra half plane of atoms that lies above (⊥) or below (⊤) the slip plane and moves parallel to it. Figure 12.11 is an idealised atomistic picture of an edge dislocation with the extra half plane of atoms above the slip plane. The displacement b is called the Burgers vector. It is constant along the dislocation line, both for edge and screw dislocations. From figure 12.10 it follows that b is perpendicular to an edge dislocation line and parallel to a screw dislocation line.

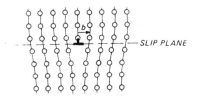

Figure 12.11. Idealised representation of an edge dislocation.

A screw dislocation does not have an extra half plane of atoms associated with it. The representation of screw dislocations is more difficult than that for an edge dislocation: a simplified example is shown in figure 12.12.

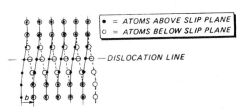

Figure 12.12. A screw dislocation in a simple cubic lattice. After reference 1.

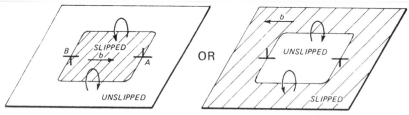

Figure 12.13. Identical dislocation lines (loops) bounding slipped and unslipped areas.

As mentioned earlier, most dislocation lines are neither pure edge nor pure screw. Furthermore, the boundary between slipped and unslipped regions often takes the form of an enclosed loop, which must be partly edge and partly screw. Such loops are illustrated in idealised form in figure 12.13. These loops are identical, and show that the relation between a dislocation and its Burgers vector is not entirely unique. The Burgers vector is ambiguous in sign unless the slipped region is specified. Also note that the screw and edge components of each loop are of opposite sign. This is because b is everywhere the same for each loop, while the screw and edge components move in opposite directions to each other. For example, the material at A is compressed (extra half plane of atoms above the slip plane, ⊥) but at B the material is extended (missing a half plane of atoms above the slip plane, which is equivalent to an extra half plane of atoms below the slip plane, ⊤).

There are stress fields associated with dislocations, and they behave as if subject to a line tension. This tension contracts a dislocation line and may collapse a loop unless there is an applied stress large enough to keep it open or even expand it, thereby causing slip.

The stress required to move a dislocation depends on a number of factors, including its character (edge and screw components), its detailed configuration in the crystal lattice, the presence of other dislocations, each with its own stress field, and barriers such as intermetallic particles and precipitates which do not fit into the crystal lattice. In polycrystalline materials grain boundaries are major barriers to dislocation movement. This is because the slip directions in neighbouring grains are generally different, and then a dislocation cannot pass from one grain to another without changing its Burgers vector, which is a high energy process.

Because there are barriers to dislocation movement, the spreading of slip within a grain and from grain to grain requires either that the dislocations cut through or bypass the barriers, or else that other dislocations are activated. In fact, both processes can occur within a grain, but it is the activation of other dislocations that enables slip to spread from grain to grain. The way in which this occurs is shown in figure 12.14. Dislocations nucleated by some distant source (usually another dislocation) in a slip plane of grain A are obstructed by the grain boundary, resulting in a dislocation pile-up. The pile-up pushes the lead dislocation hard against the boundary, such that there is a high stress concentration. Eventually the concentration of stress is sufficient to activate a dislocation source for slip in grain B, which plastically deforms to alleviate the stress concentration.

Figure 12.14. Spreading of slip from one grain to another.

The foregoing discussion is a very brief introduction to the concept of dislocations. For readers interested in a more extensive treatment there are several excellent books, including the classic by Cottrell, reference 1 of the bibliography to this chapter.

12.4. Ductile Transgranular Fracture by Microvoid Coalescence

Ductile fracture is caused by overload and, depending on the constraint, can often be recognised immediately from macroscopic examination of a failed specimen or component. If there is very little constraint there will be a significant amount of contraction before failure occurs. Figure 12.15 shows two fractured specimens after Charpy Impact testing. One failed in a ductile manner, the other was brittle. The difference is easily observed owing to the lack of constraint in this type of specimen.

Figure 12.15. Ductile and brittle fractures of impact test specimens.

However, when there is high constraint (e.g. thick sections) a ductile fracture may occur without noticeable contraction. In such cases the only macroscopic difference is the reflectivity of the fracture surface, which tends to be dull for a ductile fracture and shiny and faceted for a brittle fracture.

On a microscopic scale most structural materials fail by a process known as microvoid coalescense, which results in a dimpled appeance on the fracture surface. An example is given in figure 12.16, which shows both small and large dimples.

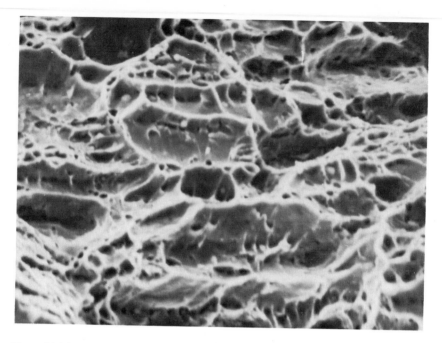

Figure 12.16. Microvoid coalescence (dimpled rupture) in a structural steel. SEM fractograph, × 1600.

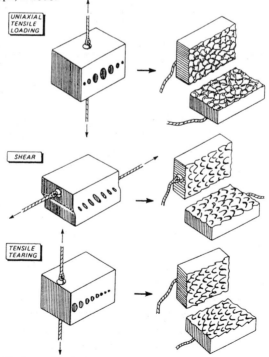

Figure 12.17. Dimple formation owing to uniaxial tensile loading, shear and tensile tearing.

Dimple shape is strongly influenced by the type of loading. This is illustrated in figure 12.17. Fracture under local uniaxial tensile loading usually results in formation of equiaxed dimples. Failures caused by shear will produce elongated or parabolic shaped dimples that point in opposite directions on matching fracture surfaces. And tensile tearing produces elongated dimples that point in the same direction on matching fracture surfaces.

The microvoids that form dimples nucleate at various internal discontinuities, the most important of which are intermetallic particles and precipitates and grain boundaries. As the local stress increases the microvoids grow, coalesce and eventually form a continuous fracture surface. A schematic of how a crack extends by microvoid formation at particles is given in figure 12.18. Note that voids can initiate both at matrix/particle interfaces and as a result of particle fracture. Also, large particles are nucleation sites for large voids, and small particles result in small voids.

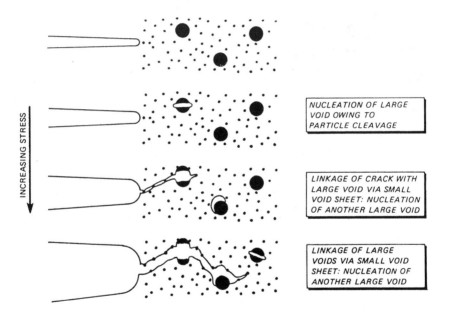

Figure 12.18. Schematic of crack extension by transgranular microvoid coalescence.

There are several dislocation models for void nucleation and growth. A model by Broek (reference 2) is illustrated in figure 12.19. The first step is the generation of dislocation loops around particles. The edge and screw components of such loops are of opposite sign. This is obvious for the screw components, which surround the particle by moving in opposite directions. However, it is less obvious for the edge components. The reasoning is as follows. Being of opposite sign the screw components will attract each other, and it is possible for them to link up and reform the original edge dislocation (⊥, extra half plane of atoms above the slip plane) if a segment

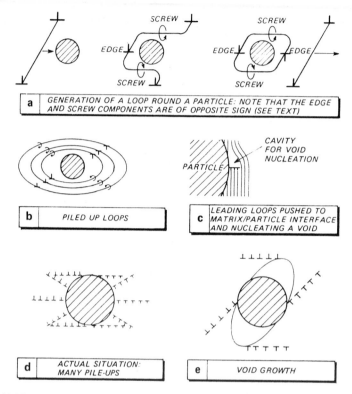

Figure 12.19. Broek's model for void nucleation and growth.

of edge dislocation of opposite sign (⊤, extra half plane of atoms below the slip plane) is left to complete the loop.

As the number of piled up loops increases, the leading loops are pushed to the matrix/particle interface and a void is nucleated. Since there are actually many pile-ups on different slip planes, once a void is nucleated it can grow by assimilating dislocations from these pile-ups.

Quantitative analysis of dislocation models of void nucleation and growth is an extremely difficult problem. Drastic simplifications are usually necessary, so that the usefulness of the models is very restricted.

12.5. Brittle Transgranular Fracture (Cleavage)

A truly brittle fracture is caused by cleavage. The term brittle fracture can be misleading, since essentially ductile fracture (microvoid coalescence) under high constraint may show the same lack of contraction expected for cleavage.

Cleavage generally takes place by the separation of atomic bonds along well-defined crystal planes. Ideally, a cleavage fracture would have perfectly matching faces and be completely flat and featureless. However, structural materials are characteristically polycrystalline with the grains more or less randomly oriented with respect to each other. Thus cleavage propagating through one grain will probably have to change direction as it crosses a grain or subgrain boundary (subgrains are regions within a grain that differ slightly in crystal orientation). Such changes in direction resulted in the faceted fracture surface shown in figure 12.15.

In addition, most structural materials contain particles, precipitates or other imperfections that further complicate the fracture path, so that truly featureless cleavage is rare, even within a single grain or subgrain. The changes of orientation between grains and subgrains and the various imperfections produce markings on the fracture surface that are characteristically associated with cleavage.

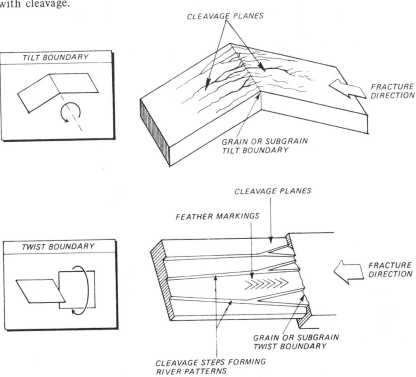

Figure 12.20. Typical features associated with cleavage.

Figure 12.20 illustrates some typical features associated with cleavage. A principal feature is river patterns, which are steps between cleavage on parallel planes. River patterns always converge in the direction of local crack propagation. If the grains or subgrains are connected by a tilt boundary, which means that they are misoriented about a common axis, the river patterns are continuous across the boundary. But if adjacent grains or subgrains are axially misoriented, i.e. they are connected by a twist boundary, the river patterns do not cross the boundary but originate at it. Besides river patterns a distinct feature of cleavage is feather markings. The apex of these fan-like markings points back to the fracture origin, and therefore this feature can also be used to determine the local direction of crack propagation.

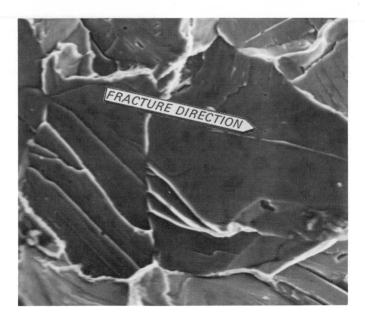

Figure 12.21. Cleavage in BS4360 structural steel. SEM fractograph, ×900.

Cleavage occurs in a number of materials, but it is especially important because it occurs in structural steels. An example is shown in figure 12.21.

Knott's book on fracture mechanics (reference 3) reviews dislocation models of cleavage with particular attention to steels, including a model by Smith (reference 4) that incorporates the important microstructural feature of grain boundary carbides, which are known to greatly influence the fracture toughness of steels.

Smith's model is shown in figure 12.22. As a consequence of a tensile stress σ a brittle intergranular carbide is subjected to a concentrated shear

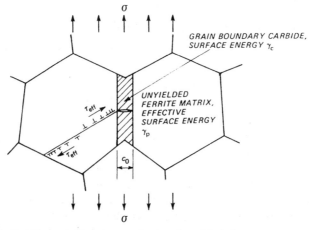

Figure 12.22. Smith's model for cleavage fracture in mild steel.

stress ahead of a dislocation pile-up of length d, the average grain size. The effective shear stress, τ_{eff}, is the maximum stress that can be attained before yielding occurs, i.e. before the stress concentration due to the pile-up is relieved by the spreading of slip to a neighbouring grain. The condition for cracking the carbide is

$$\tau_{eff} \geqslant \sqrt{\frac{4E\gamma_c}{(1 - \nu^2)\pi d}} .$$ (12.1)

Once the carbide is cracked there are two possibilities:

(1)　τ_{eff} is high enough to propagate the carbide crack into the ferrite matrix, that is: cleavage of the matrix is nucleation controlled. For this to happen the condition is

$$\tau_{eff} \geqslant \sqrt{\frac{4E\gamma_p}{(1 - \nu^2)\pi d}} .$$ (12.2)

(2)　τ_{eff} at yielding lies between the limits set by equations (12.1) and (12.2). This is much more likely. In this case the carbide thickness, c_o, gives the size of initial crack in a Griffith-type energy balance criterion for crack extension:

$$\sigma_f^2(\frac{c_o}{d}) + \tau_{eff}^2 \left[1 + \frac{4}{\pi}(\frac{c_o}{d})^{1/2} \frac{\tau_i}{\tau_{eff}}\right]^2 \geqslant \frac{4E\gamma_p}{(1 - \nu^2)\pi d}$$ (12.3)

where σ_f is the critical stress for cleavage fracture and τ_i is a so-called friction stress, which includes a number of factors that cause the crystal lattice to resist dislocation movement.

Equation (12.3) shows that larger carbides should result in a lower cleavage fracture stress, as is observed. The model has been tested quantitatively, and gives good predictions of σ_f for steels with large grain sizes. However, it is difficult to check the model for fine grained steels, since fine grains are invariably associated with thin carbides, i.e. carbide size variation is not possible.

12.6. Transgranular Fracture by Fatigue

Fatigue results in very distinctive fracture appearances. In general there are three different features that can be observed:

- the location of fatigue initiation
- the fracture surface resulting from fatigue crack growth
- the final fracture due to overload.

Each of these features can give specific information about the fatigue process in a structural component.

Fatigue initiation nearly always occurs at an external surface, though there are important exceptions like jet engine compressor discs where internal fatigue origins have been found. Careful examination of a fatigue initiation site may reveal the cause of fatigue, for example stress concentrations due to a notch, machining grooves or corrosion pitting. This kind of information

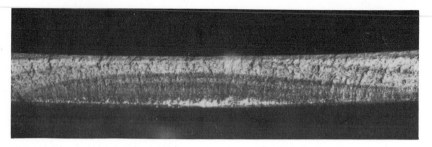

Figure 12.23. Beach markings on a fatigue fracture surface in a thin walled pipe.
Optical fractograph, ×5.

is of primary importance for analysis of service failures but is less relevant
to fracture mechanics studies, although in recent years the problem of short
crack growth at notches (see section 9.8) has received widespread attention.

Fatigue fracture surfaces tend to be macroscopically flat and smooth, and
will often show 'beach markings', which occur owing to variations in the load
history. An example is given in figure 12.23.

Beach markings can be very useful, since they give information about the
shapes of the fatigue crack front at various stages of growth, and these shapes
are diagnostic for the type of fatigue loading. Some typical beach markings

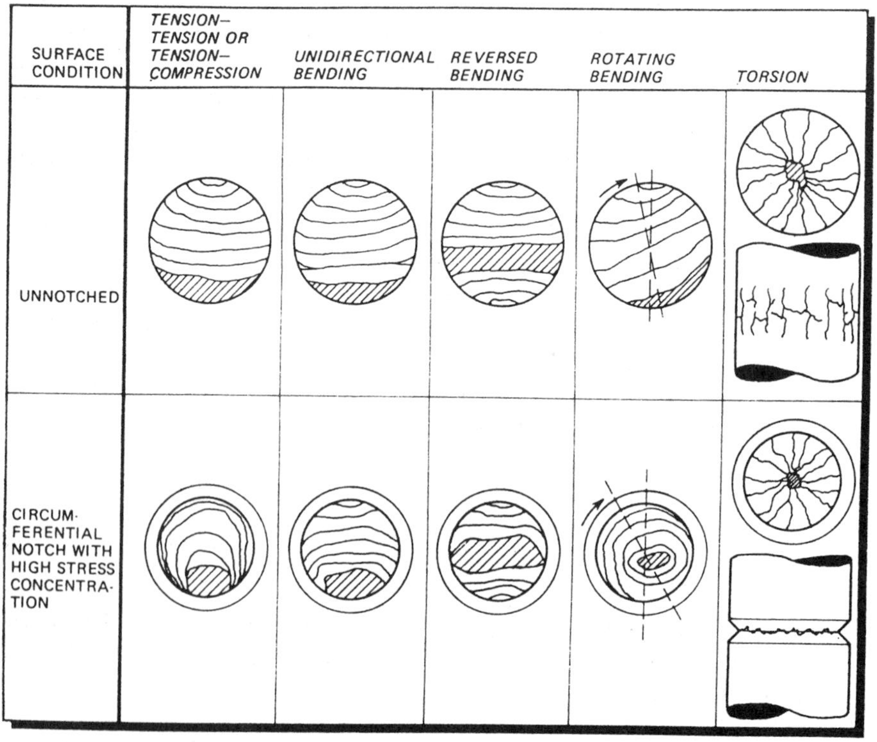

SURFACE CONDITION	TENSION–TENSION OR TENSION–COMPRESSION	UNIDIRECTIONAL BENDING	REVERSED BENDING	ROTATING BENDING	TORSION
UNNOTCHED					
CIRCUM-FERENTIAL NOTCH WITH HIGH STRESS CONCENTRA-TION					

Figure 12.24. Examples of beach markings for different types of loading of a cylindrical
bar. ▨ = area of final fracture. After reference 5.

for different types of loading imposed on a cylindrical bar are indicated schematically in figure 12.24. Many more examples are presented in reference 5 for both cylindrical and rectangular cross-sections.

The area of final fracture (shown hatched in figure 12.24) gives an indication of the magnitude of the loads. A large final fracture area indicates that K_{Ic} or K_c was exceeded at a relatively short crack length, which means that either the maximum load was high or the fracture toughness low, or both. Knowledge of a material's fracture toughness removes the ambiguity, and providing the crack geometry is not too complex a fair estimation of the maximum stress at failure can be obtained from an approximate expression for the stress intensity factor.

On a microscopic scale the most characteristic features of fatigue are the striations that occur during region II (continuum mode) crack growth, see figure 9.3. The striations represent successive positions of the crack front. Each striation is formed during one load cycle but, especially under variable amplitude loading, not every load cycle need result in a striation.

Figure 12.25 shows two different examples of fatigue striations. Aluminium alloys generally give well-defined regular striations, but steels do not. Besides the influence of type of material, the environment also has a strong effect on striation appearance. Note that striations are perpendicular to the local direction of crack growth. This is sometimes helpful in tracing crack growth backwards in order to determine the exact location of fatigue initiation.

Since each striation is formed during one load cycle the spacing between striations is an indication of local crack growth rates, particularly if it is evident from their regularity that they have been formed by constant amplitude loading (typical examples are propellers, helicopter rotor blades and rotating components in power generating equipment). These local crack growth rates can be used to determine the fatigue stresses provided the following procedure is possible:

(1) Assume the local striation spacing is equal to da/dn and read off values of ΔK from da/dn–ΔK crack growth rate curves for different stress ratios, R.

(2) From the crack length and geometry calculate the corresponding maximum and minimum stresses, and hence the mean stress, for each R value.

(3) Make an independent estimate of the mean stress from design and operating data.

(4) Select or estimate the most appropriate R value and hence the most likely values of the fatigue stresses.

Micromechanistic modelling of fatigue is a subject of considerable interest and speculation, especially concerning fatigue striations. There does, however, appear to be a fairly consistent picture of fatigue crack initiation at an external surface.

Under fatigue loading the surface material tends to deform by cyclic slip concentrated in so-called persistent slip bands (PSBs) with irregular notched

2024-T3 ALUMINIUM ALLOY

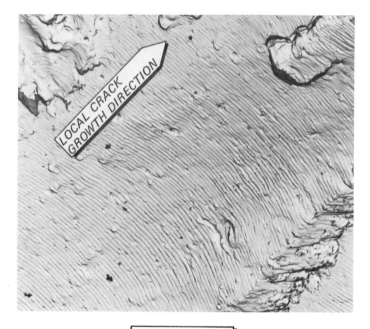

ST E460 STEEL

Figure 12.25. Fatigue striations for an aluminium alloy and a structural steel tested in normal air. TEM replica fractographs, ×4000.

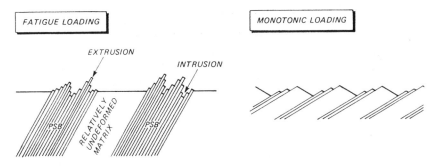

Figure 12.26. Illustration of slip during fatigue and monotonic loading.

profiles consisting of extrusions and intrusions. As figure 12.26 shows, this slip distribution is quite different from that produced by monotonic loading.

Continuing cyclic slip leads to deepening of the intrusions and eventually the formation of a crack along the slip plane. This slip plane cracking may extend a few grain diameters into the material, but then changes to a continuum mechanism of crack propagation.

A simple dislocation model for the formation of extrusions and intrusions is shown in figure 12.27. The dislocations pile up at a grain boundary and tend to form pairs (dipoles) owing to interaction of their stress fields. An

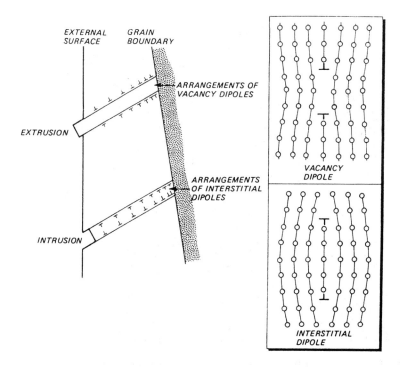

Figure 12.27. Simple dislocation model of extrusions and intrusions. After reference 6.

arrangement of vacancy dipoles means that the material between them has fewer half planes of atoms than the matrix: these half planes have been transported beyond the surface to form an extrusion. On the other hand, an arrangement of interstitial dipoles means that the material between them has more half planes of atoms than the matrix, and this results in an intrusion.

The dislocation arrangements within a PSB are actually much more complicated than those assumed by the model in figure 12.27. Nevertheless, analysis of the model predicted the form of observed relations between fatigue initiation life and cyclic plastic strain and between grain size and fatigue strength in some materials, reference 6.

Various models of fatigue striation formation have been proposed. Most consider only plastic flow at the crack tip and ignore the potential contribution of the environment. One such model, the plastic blunting process, is shown in figure 12.28. During uploading shear deformation concentrates at first at 'ears' on either side of the crack tip. Later the crack tip itself advances and blunts. During unloading the shear deformations are such that a new pair of ears is formed at the crack tip, thereby producing the characteristic striation marking on the fracture surface. Why this ear formation should happen during unloading has never been explained. However, it does appear to occur in both inert and aggressive environments, reference 8.

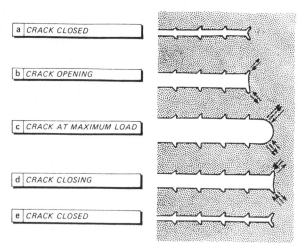

a CRACK CLOSED

b CRACK OPENING

c CRACK AT MAXIMUM LOAD

d CRACK CLOSING

e CRACK CLOSED

Figure 12.28. The plastic blunting model of fatigue striation formation. After reference 7.

12.7. Intergranular Fracture

Intergranular fractures are typically the result of sustained load fracture, discussed in section 12.8, or a lack of ductility in the material owing to segregation of embrittling elements and particles and precipitates to the grain boundaries, for instance in temper-embrittled steels and overaged Al-Zn-Mg-Cu aluminium alloys.

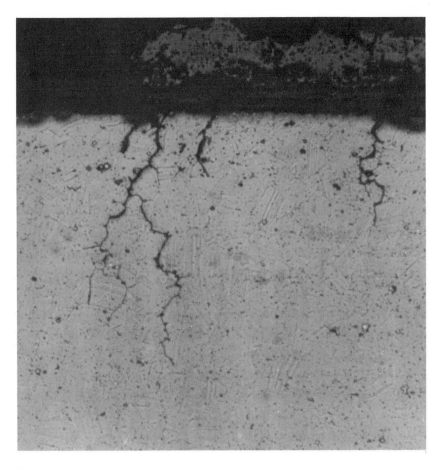

Figure 12.29. Intergranular cracks in Inconel 625 (a nickel-base alloy for high temperature use). Optical metallograph, ×320.

It is not possible to distinguish macroscopically between intergranular
fracture and brittle transgranular fracture: both appear faceted. However,
metallographic cross-sections through fracture surfaces and cracks will show
whether the fracture path is intergranular, see for instance figure 12.29.

There are two main types of intergranular fracture appearance:

(1)   Grain boundary separation with microvoid coalescence. This type
      of intergranular fracture occurs during overload failure of some
      steels and aluminium alloys, and also other materials.

(2)   Grain boundary separation without microvoid coalescence. This
      type of intergranular fracture occurs during overload failure of
      temper-embrittled steels and refractory metals like tungsten, and
      also during sustained load fracture (creep, stress corrosion cracking,
      embrittlement by hydrogen and liquid metals).

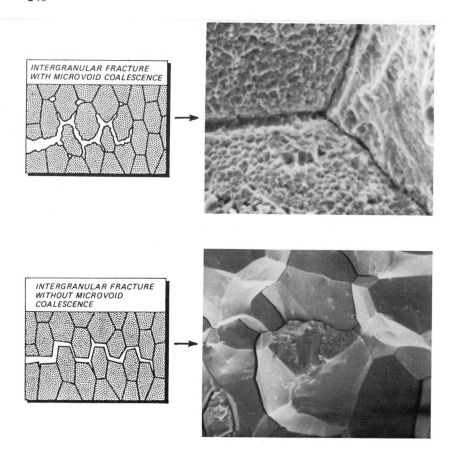

Figure 12.30. Intergranular fracture with and without microvoid coalescence. SEM fractographs, top ×1400, bottom ×500.

Examples of both types are given in figure 12.30. The dimples on the grain boundary facets are the main distinguishing feature of intergranular fracture with microvoid coalescence.

Intergranular fractures are not always readily identifiable. Figure 12.31 shows schematically an intergranular fracture along flat elongated grains, which often occur in rolled sheet and plate materials as a consequence of mechanical working. This type of intergranular fracture exhibits few grain boundary junctions and is relatively featureless.

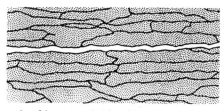

Figure 12.31. Schematic of intergranular fracture along elongated grains.

12.8. Types of Sustained Load Fracture

In this section we shall first consider the characteristics of creep fracture, which is primarily a plasticity-induced mechanism of fracture at elevated temperatures, and then some examples of sustained load fracture induced by aggressive environments.

Creep Fracture

At temperatures in excess of $0.5\ T_m$ (the melting point of a material) time-dependent deformation and rupture (creep) is a primary design consideration. In many applications, such as gas turbines and steam boilers, the operating temperature is limited by the creep characteristics of materials.

Creep fractures in most commercial alloys are intergranular. There are two forms of intergranular separation, depending on the load and temperature:

(1) At high loads and low temperatures in the creep range the fractures tend to originate at grain boundary junctions (triple points), rather than on the boundaries.

(2) At lower loads and higher temperatures (the typical creep situation) fracture results more from the formation of voids along grain boundaries, especially those boundaries perpendicular to the loading direction. This process is called cavitation.

The initiation of both types of creep fracture is illustrated schematically in figure 12.32. The fracture surface of a high load—low temperature creep fracture consists of intergranular facets without microvoid coalescence, and is similar in appearance to the lower fractograph in figure 12.30.

On the other hand, the fracture surface of a low load—higher temperature fracture will often exhibit voids on the grain boundary facets. These voids

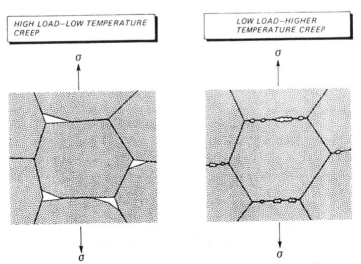

Figure 12.32. Schematic of the two main forms of creep fracture initiation.

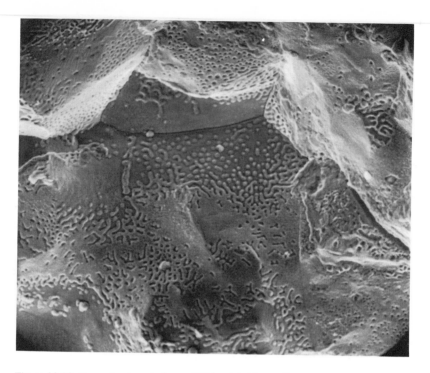

Figure 12.33. Creep fracture in Inconel W (a nickel-base alloy) showing distinctive patterns of void coalescence on grain boundary facets. SEM fractograph, ×400.

can be observed from a metallographic section, or even by optical microscopy of a replica from a polished surface. (Surface replicas are widely used to assess creep damage in large structures such as steam pipes in power generating plants.) The grain boundary voids can coalesce to resemble dimpled rupture, as in the upper fractograph in figure 12.30, but they can also coalesce in very distinctive patterns. An example of such patterns is shown in figure 12.33.

Nucleation of creep voids most probably occurs by a combination of grain boundary sliding, in which the grain boundary behaves like a slip plane, and stress-assisted diffusion and agglomeration of lattice vacancies. At triple points grain boundary sliding results in geometric incompatibilities (and hence stress concentrations) which can be accommodated by vacancy diffusion to nucleate voids. On the grain boundaries the voids are nucleated by vacancy diffusion, especially to matrix particle interfaces, and grain boundary sliding can assist this process.

There is less certainty as to the controlling mechanism of void growth and coalescence to fracture, which is a very complicated process. A fairly recent attempt to resolve the inconsistencies in earlier creep fracture models is due to Edward and Ashby, reference 9. They proposed that void growth on cavitated grain boundaries occurs by vacancy diffusion at a rate controlled not by the stress but by deformation of uncavitated material surrounding the voids.

Sustained Load Fracture in Aggressive Environments

Important kinds of sustained load fracture in aggressive environments include:

- stress corrosion cracking
- cracking due to embrittlement by internal or external (gaseous) hydrogen
- liquid metal embrittlement.

The term *stress corrosion cracking* covers a very wide range of material–environment interactions, and it is not possible to give more than a brief overview of some proposed fracture mechanisms here. For the interested reader a good background to the subject is provided by references 10–13 of the bibliography to this chapter.

Stress corrosion fractures can be transgranular or intergranular. Sometimes they are a mixture, though one mode usually predominates. Macroscopically transgranular stress corrosion cracks are often faceted. On a microscale the fracture surface may show a feather-shaped appearance as in figure 12.34 or can strongly resemble mechanical cleavage (see figure 12.21) as in aqueous stress corrosion of titanium and magnesium alloys.

In fact, the cleavage-like appearance has led to models of environment-induced cleavage owing to adsorption of specific ions, for instance hydrogen, at the crack tip or to hydrogen absorption followed by internal decohesion which links up with the main crack.

Figure 12.34. Feather-shaped crack surface of a transgranular stress corrosion crack in austenitic stainless steel. SEM fractograph, ×500.

TYPE OF FAILURE	FILM RUPTURE MODEL	STRESS-ASSISTED INTER-GRANULAR CORROSION	TUNNEL MODEL	ADSORPTION (INCLUDING HYDROGEN)	HYDROGEN ABSORPTION + DECOHESION
transgranular cracking of austenitic stainless steels	•		•	•	•
intergranular cracking of low strength ferritic steels	•	•		•	
intergranular cracking of austenitic stainless steels	•	•		•	
intergranular cracking of high strength low alloy steels				•	•
transgranular cracking of low strength ferritic steels	•			•	•
intergranular cracking of titanium alloys	•	•		•	
transgranular cracking of titanium alloys	•			•	•
intergranular cracking of aluminium alloys	•			•	•

Figure 12.35. Classification of some types of stress corrosion and models suggested to explain them.

A classification of stress corrosion cracking models proposed for some structural materials is given in figure 12.35. The fact that different models have been suggested to explain each type of failure is an indication of the complexity of stress corrosion cracking. A short description of each model follows.

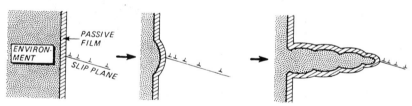

Figure 12.36. The film rupture model of stress corrosion cracking.

The *film rupture model* is also sometimes called the slip dissolution model. The model is illustrated in figure 12.36. Emergent slip bands at a surface or crack tip break a passive film and the crack propagates owing to local dissolution of metal.

The *stress-assisted intergranular corrosion model* is sometimes called the brittle film mechanism. It is shown in figure 12.37. The model requires the environment to produce a mechanically weak surface film that grows preferentially along grain boundaries and eventually cracks under stress. The crack is arrested by plastic deformation of the metal, which then reacts with the environment to reform the film, and so on.

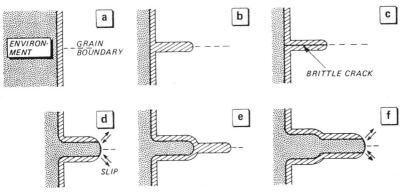

Figure 12.37. The stress-assisted intergranular corrosion model.

The *tunnel model* was proposed specifically for transgranular cracking of austenitic stainless steels in order to explain the typical feather-shaped appearance shown in figure 12.34. It involves the formation of arrays of corrosion tunnels at slip steps as depicted in figure 12.38. As the tunnels grow the ligaments of metal between them become more highly stressed and eventually fail by ductile rupture. The process is then repeated.

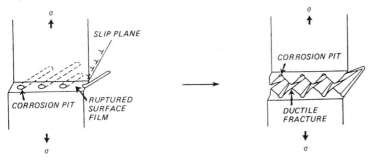

Figure 12.38. The tunnel model of stress corrosion cracking.

The *adsorption model* is a general one proposing that an environmental species can interact with strained crystal lattice bonds at the crack tip to cause a reduction in bond strength and hence brittle crack extension. There are many objections to the applicability of this model for stress corrosion. It has enjoyed some popularity for explaining embrittlement by gaseous hydrogen and liquid metals, but even these possibilities now appear to be unlikely, as will become evident from the remainder of this section.

The *hydrogen absorption + decohesion theory* of stress corrosion cracking involves stress-assisted hydrogen diffusion to a location ahead of the crack tip. The increased hydrogen concentration at this location then results in cracking that links up with the main crack. This model, or a modification of it, may be valid for a number of material—environment combinations. In the next chapter, section 13.4, it is shown that the model can be used to quantitatively predict stress corrosion crack growth rates for high strength steels.

Embrittlement by internal or external hydrogen also occurs in a wide range of materials. It is particularly important for steels, and also for high strength nickel-base alloys, titanium alloys and materials used in the nuclear power industry (zirconium, hafnium, niobium, uranium).

Most hydrogen embrittlement fractures are intergranular, but cleavage-like cracking can also occur, especially if transgranular brittle hydrides form and act as crack nucleation sites. Besides hydride formation in the material, hydrogen can remain in the lattice and interact with dislocations and other lattice defects, including segregation to matrix/particle interfaces. Hydrogen may also react to give surface hydrides that lower the fracture stress at crack tips, or else these surface hydrides form brittle films along grain boundaries such that the type of mechanism shown in figure 12.37 applies. In short, hydrogen embrittlement is no one thing. For further reading references 13–15 may be consulted.

Because of its importance much attention has been paid to quantitative analyses of hydrogen embrittlement in high strength steels. As in the case of stress corrosion these analyses will be discussed in the next chapter, section 13.4. The results of the analyses indicate that both external and internal hydrogen embrittlement are caused by lattice decohesion ahead of the main crack, in a similar manner to stress corrosion cracking of high strength steels.

The final topic in this section is *liquid metal embrittlement*. Again this is a widespread phenomenon. It is usually of much less practical significance than stress corrosion cracking and hydrogen embrittlement, but the most recently proposed mechanism is very interesting since it could have more general applications.

Liquid metal embrittlement results in drastic losses in macroscopic ductility. The fracture path can be intergranular or transgranular, and consists of facets that appear very brittle at low and intermediate magnifications (up to $\sim \times 1000$). Because of this, it was for a long time thought that liquid metal embrittlement was due to adsorption of liquid metal atoms at the crack tip and a consequent reduction in crystal lattice bond strength so that brittle crack extension occurs. The model proposed on this basis is shown in figure 12.39. (This model is essentially the same as that suggested, and now largely

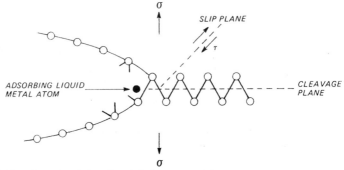

Figure 12.39. Adsorption-induced embrittlement model.

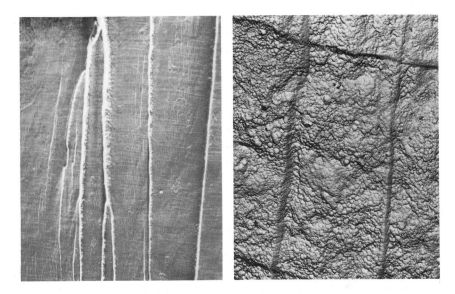

Figure 12.40. Liquid metal embrittlement of an aluminium single crystal. Left: SEM fractograph, ×50. Right: TEM replica fractograph, ×5000.
Courtesy S.P. Lynch.

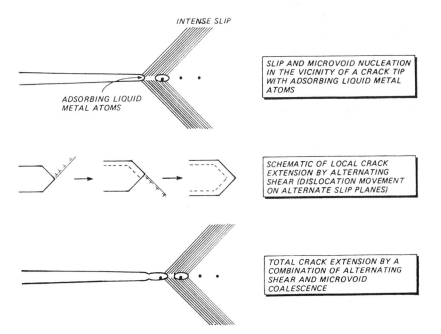

INTENSE SLIP

ADSORBING LIQUID
METAL ATOMS

SLIP AND MICROVOID NUCLEATION IN THE VICINITY OF A CRACK TIP WITH ADSORBING LIQUID METAL ATOMS

SCHEMATIC OF LOCAL CRACK EXTENSION BY ALTERNATING SHEAR (DISLOCATION MOVEMENT ON ALTERNATE SLIP PLANES)

TOTAL CRACK EXTENSION BY A COMBINATION OF ALTERNATING SHEAR AND MICROVOID COALESCENCE

Figure 12.41. Crack extension in liquid metals owing to adsorption-induced intense shear and microvoid coalescence.

discounted, to account for stress corrosion cracking and hydrogen embrittlement.) The hypothesis is that adsorbing atoms lower the tensile stress, σ, required to break lattice bonds, but do not influence the shear stress, τ, necessary to move dislocations to cause plastic deformation and crack tip blunting.

Recently, however, Lynch (references 16, 17) has shown that the apparently brittle fractures caused by liquid metal embrittlement are in most cases covered by shallow microvoids. An example is given in figure 12.40. This shows macroscopically brittle 'cleavage' in an aluminium single crystal and a TEM replica fractograph of part of the fracture surface. The microscopic ductility is considerable. In view of this microscopic ductility Lynch proposed that adsorbing liquid metal atoms lower both the tensile and shear stresses at a crack tip. Lowering the shear stress facilitates dislocation movement near the crack tip, and the overall effect is a highly localised concentration of slip and hence much less total plastic deformation during crack extension.

As visualised by Lynch, actual crack extension generally occurs by a combination of intense local shear and microvoid formation and coalescence with the main crack, see figure 12.41. This mechanism, with or without microvoid coalescence, may be applicable to other environments besides liquid metals, and to environmental fatigue crack growth as well as sustained load fracture.

12.9. Bibliography

1. Cottrell, A.H., Dislocations and Plastic Flow in Crystals, Oxford University Press (1953): London.
2. Broek, D., *A Study on Ductile Fracture*, Ph.D. Thesis, Delft University of Technology, The Netherlands (1971).
3. Knott, J.F., Fundamentals of Fracture Mechanics, Butterworths (1973): London.
4. Smith, E., *The Nucleation and Growth of Cleavage Microcracks in Mild Steel*, Proceedings of the Conference on the Physical Basis of Yield and Fracture, Institute of Physics and Physical Society, pp. 36–46 (1966): Oxford.
5. Metals Handbook 8th Edition, Vol. 9, Fractography and Atlas of Fractographs, American Society for Metals (1974): Metals Park, Ohio.
6. Tanaka, K. and Mura, T., *A Dislocation Model for Fatigue Crack Initiation*, Journal of Applied Mechanics, Vol. 48, pp. 97–103 (1981).
7. Laird, C., *The Influence of Metallurgical Structure on the Mechanisms of Fatigue Crack Propagation*, Fatigue Crack Propagation, ASTM STP 415, pp. 131–168 (1967): Philadelphia.
8. Lynch, S.P., *Mechanisms of Fatigue and Environmentally Assisted Fatigue*, Fatigue Mechanisms, ASTM STP 675, pp. 174–213 (1979): Philadelphia.
9. Edward, G.H. and Ashby, M.F., *Intergranular Fracture during Power-Law Creep*, Acta Metallurgica, Vol. 27, pp. 1505–1518 (1979).
10. Logan, H.L., The Stress Corrosion of Metals, John Wiley and Sons, Inc. (1966): New York.
11. Proceedings of the Conference on the Fundamental Aspects of Stress Corrosion Cracking, National Association of Corrosion Engineers (1969): Houston, Texas.
12. The Theory of Stress Corrosion Cracking in Alloys, North Atlantic Treaty Organization Scientific Affairs Division (1971): Brussels.
13. Stress Corrosion Cracking and Hydrogen Embrittlement of Iron Base Alloys, National Association of Corrosion Engineers (1977): Houston, Texas.
14. Hydrogen in Metals, American Society for Metals (1974): Metals Park, Ohio.

15. Effect of Hydrogen on Behaviour of Materials, Metallurgical Society of AIME (1976): New York.

16. Lynch, S.P., *The Mechanism of Liquid-Metal Embrittlement-Crack Growth in Aluminium Single Crystals and other Metals in Liquid-Metal Environments*, Aeronautical Research Laboratories Materials Report 102 (1977): Melbourne, Australia.

17. Lynch, S.P., *Metallographic and Fractographic Aspects of Liquid-Metal Embrittlement*, Environmental Degradation of Engineering Materials in Aggressive Environments, Virginia Polytechnic Institute, pp. 229–244 (1981): Blacksburg, Virginia.

13. THE INFLUENCE OF MATERIAL BEHAVIOUR ON FRACTURE MECHANICS PROPERTIES

13.1. Introduction

In this final chapter we shall provide some insight into the ways in which the actual behaviour of materials influences the fracture mechanics parameters used to describe crack extension. Most of the discussion concerns LEFM parameters, mainly because they are more widely used. Figure 13.1 lists the various types of fracture considered in this course and the relevant fracture mechanics parameters. At present a significant contribution by EPFM is made only in the case of fracture toughness and slow stable tearing.

| TYPE OF FRACTURE | FRACTURE MECHANICS | |
	• CONDITIONS	• PARAMETERS
• INITIATION OF UNSTABLE CRACK GROWTH (FRACTURE TOUGHNESS)	LEFM	K_{Ic}, K_c
	EPFM	J_{Ic}, COD
• SLOW STABLE TEARING	LEFM	K_R
	EPFM	R-Line, T
• FATIGUE CRACK GROWTH	LEFM	$\Delta K_{th}, \Delta K$
	EPFM	$\Delta K_e, \Delta J$
• SUSTAINED LOAD FRACTURE (EXCLUDING CREEP)	LEFM	K_{Ith}, K_I, K_{Imax}
• DYNAMIC FRACTURE	LEFM	K_{Id}, K_{Im}

Figure 13.1. Types of fracture and the parameters for describing them.

The influence of material behaviour on fracture mechanics characterization of crack extension will be described in sections 13.2−13.4 as follows:

(13.2) The effects of crack tip geometry:
- blunting
- branching and kinking
- through-thickness irregularity
- change of mode.

(13.3) The effects of fracture path and microstructure:
- transgranular and intergranular fracture; mixed fracture paths
- effects of microstructure
  - second phases
  - particles and precipitates
  - grain size
  - fibering and texturing owing to mechanical working.

(13.4) Fracture mechanics and the mechanisms of fracture:
- fracture by microvoid coalescence
- cleavage in steels
- fatigue crack growth

- sustained load fracture
- superposition or competition of sustained load fracture and fatigue.

This subdivision is convenient in that the subjects are presented more or less in the order of increasing complexity. However, they are often strongly interrelated, as will be seen.

13.2. The Effects of Crack Tip Geometry

LEFM analysis begins with slit-shaped, unbranched cracks of zero tip radius. The latter assumption is obviously unrealistic, owing to the occurrence of plasticity which blunts the crack. Also, most real cracks exhibit at least small amounts of branching, kinking and through-thickness irregularity. These problems have been mentioned in section 10.4.

Changes in mode of loading can occur. The most important are the mode II → mode I and mode I → combined modes I and II transitions during fatigue crack propagation at low stress intensity levels.

Crack Tip Blunting

Crack tip blunting always occurs in practice. The blunting may be very limited, e.g. during environmental fatigue crack propagation and sustained load fracture in high strength materials. In such cases sharp crack LEFM is usually adequate for characterizing crack extension.

On the other hand, blunting is very important during EPFM fracture toughness testing. This is recognised in the procedures for determining J_{Ic} (section 7.4) and COD (section 7.5).

As stated in section 2.2, the mode I elastic stress field equations for a blunt crack were derived by Creager and Paris, reference 1 of the bibliography to this chapter. The crack tip coordinate system is shown in figure 13.2.

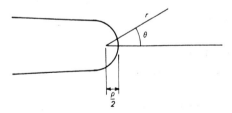

Figure 13.2. Crack with a finite tip radius, ρ.

Under mode I loading

$$\sigma_x = \frac{K_I}{\sqrt{2\pi r}} \cos\frac{\theta}{2}(1 - \sin\frac{\theta}{2}\sin\frac{3\theta}{2}) - \frac{K_I}{\sqrt{2\pi r}} \frac{\rho}{2r} \cos\frac{3\theta}{2}$$

$$\sigma_y = \frac{K_I}{\sqrt{2\pi r}} \cos\frac{\theta}{2}(1 + \sin\frac{\theta}{2}\sin\frac{3\theta}{2}) + \frac{K_I}{\sqrt{2\pi r}} \frac{\rho}{2r} \cos\frac{3\theta}{2} \qquad (13.1)$$

$$\tau_{xy} = \frac{K_I}{\sqrt{2\pi r}} \sin\frac{\theta}{2}\cos\frac{\theta}{2}\cos\frac{3\theta}{2} \qquad - \frac{K_I}{\sqrt{2\pi r}} \frac{\rho}{2r} \sin\frac{3\theta}{2}$$

For sharp cracks ρ is small. Unless one is interested in regions very close to the crack tip the terms with (ρ/r) can be neglected and equations (13.1) reduce to the standard form given in equations (2.10).

Blunting has a minor effect on the size and shape of the plastic zone, reference 2. However, for plane strain conditions the distribution of σ_y ahead of the crack tip is greatly altered, as figure 13.3 shows. A blunted crack tip acts as a free surface and locally reduces the triaxiality of the state of stress (at $r = \rho/2$ $\sigma_x = 0$ for $\theta = 0$), see section 3.5. Thus increased blunting decreases the maximum stress and moves its location away from the crack tip towards the elastic-plastic boundary. These effects have been used to assess the effects of blunting on fracture toughness and sustained load fracture, as will be discussed in section 13.4.

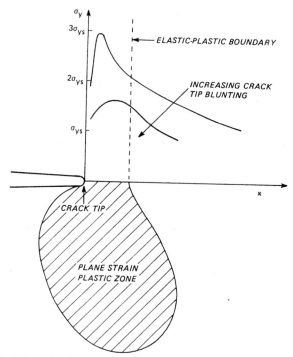

Figure 13.3. Schematic of the effect of crack tip blunting on σ_y for an elastic-perfectly plastic material under plane strain.

Crack Branching and Kinking

Microscopie crack branching and kinking commonly occur. Their significance is often overlooked when the fracture mechanics-related properties of materials are determined. Macrobranching also occurs, but only during region II sustained load fracture and dynamic fracture. The reason for this limitation is that macrobranching depends on there being little or no tendency for one branch to outrun the other(s), and this is only possible when crack growth rates are virtually independent of crack length and stress intensity,

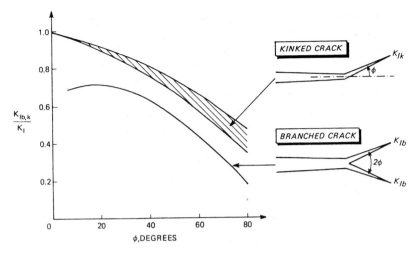

Figure 13.4. Mode I stress intensities for branched and kinked cracks. After reference 3.

as is the case for region II sustained load fracture, see figure 10.4, and for dynamic crack branching, section 11.2.

Crack branching and kinking lower the mode I stress intensities at the crack tips. This is illustrated in figure 13.4. Consequently, for nominally mode I crack extension it may be expected that microbranching and kinking will result in

- higher fracture toughness
- higher thresholds for fatigue crack growth (ΔK_{th}) and sustained load fracture (K_{Ith})
- lower crack growth rates in fatigue and region I sustained load fracture.

Through-Thickness Irregularity

Fracture mechanics analyses of through-thickness crack front irregularity are not available. Nevertheless, for both sustained load fracture (reference 4) and fatigue crack growth (references 5 and 6) it has been found that increased irregularity results in lower crack growth rates.

Change of Mode

Two types of change of mode of crack growth will be considered here:

(1) The flat-to-slant transition at high stress intensities, for both monotonic and fatigue loading.

(2) The mode II → mode I and mode I → combined modes I and II transitions during fatigue crack propagation at low stress intensities.

The flat-to-slant transition under monotonic loading was discussed in section 3.6. A similar transition occurs in fatigue, figure 13.5. Shear lips gradually develop as the fatigue crack grows and the crack front becomes increasingly curved. The fracture plane rotates from flat to slant with a component of mode III loading. As in the case of monotonic loading, this fracture

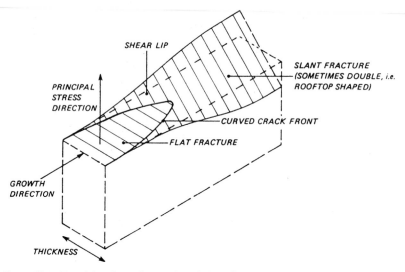

Figure 13.5. Transition from flat to slant fatigue fracture.

plane rotation is related to a change from predominantly plane strain to plane stress conditions. However, a change in stress state is not the only factor. The flat-to-slant transition is strongly dependent on the environment: more aggressive environments postpone the transition.

The flat-to-slant transition affects fatigue crack growth rates. Under constant amplitude loading the slope of the fatigue crack growth rate curve will slightly decrease. Under variable amplitude loading one or more tensile peak loads can induce a change from flat to slant fracture, which then reverts to flat fracture under continued less severe load cycling. Such crack front geometry changes are a potential complication for predicting crack growth, since re-orientation of the crack front back to flat fracture is likely to affect the amount of crack growth retardation due to the peak loads.

The mode II → mode I transition during fatigue crack propagation is of fundamental importance since it often represents the initiation and early growth of fatigue cracks under constant or increasing stress cycling, $\Delta\sigma$. Figure 13.6 gives a schematic of this process as follows:

(a) Cyclic slip begins in a surface grain and occurs mainly on one or a few sets of crystal planes.

(b) This usually leads to slip plane cracking (mode II), which results in a faceted fracture surface, and spreading of cyclic slip to an adjacent grain. Again the slip is mainly on one or a few sets of crystal planes.

(c) The second grain also cracks along the slip plane. Cyclic slip in the interior is now activated on several sets of crystal planes. This enables mode I crack extension.

(d) Cyclic slip on several sets of crystal planes results in a continuum mechanism of crack propagation, often characterized by fatigue striations.

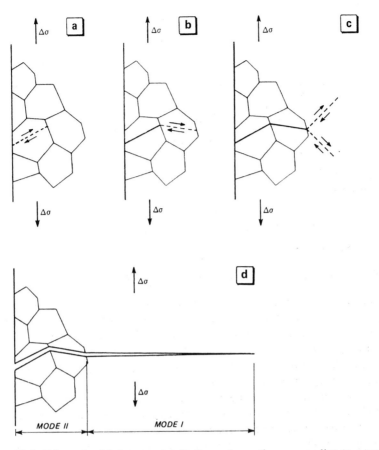

Figure 13.6. Schematic of fatigue crack initiation and growth corresponding to a transition from mode II to mode I:

(For more details on fatigue crack initiation and propagation the reader is referred to the previous chapter, section 12.6.)

The mode I → combined modes I and II transition is also important. It generally occurs when a fatigue crack is growing at low stress intensity levels either when the overall ΔK is gradually decreasing towards the threshold, reference 7, or else after a tensile peak load, reference 8. In both cases faceted fracture (not necessarily slip plane cracking) occurs and the crack path is effectively a series of kinked cracks. This means that apart from a decreasing ΔK or the effects of residual stresses due to a peak load there are two purely geometric effects that contribute to lower crack growth rates:

(1)   The crack tip mode I stress intensities for kinked cracks are lower than those for pure mode I cracks, e.g. figure 13.4.

(2)   There is an increase in crack closure owing to increased contact of the fracture surfaces. This is illustrated in figure 13.7.

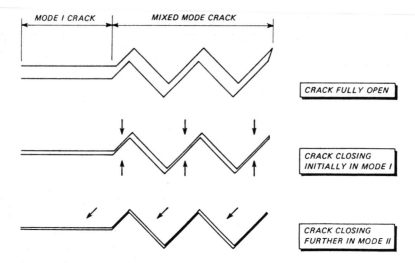

Figure 13.7. Schematic diagram of increased crack closure owing to combined mode I and mode II fatigue crack growth.

13.3. The Effects of Fracture Path and Microstructure

The effects of fracture path and microstructure on the fracture mechanics properties of materials are based on geometric and inherent characteristics. More specifically, the amounts of plastic deformation and energy required for crack extension depend on both crack tip geometry and the fracture path. In turn, the fracture path is determined by the microstructure and, sometimes, by microstructurally influenced material-environment interactions.

In this section we shall consider transgranular and intergranular fracture and mixtures of different fracture paths before proceeding to the influences of microstructure.

Transgranular Fracture

Figure 13.8 lists important kinds of transgranular fracture, their occurrence and general descriptions of their geometric and mechanical characteristics. The amount of energy required for crack extension depends on these characteristics. At one extreme microvoid coalescence, which is the usual way in which stable and unstable crack extension occur, requires relatively high amounts of energy. On the other hand cleavage is a low energy fracture.

Transgranular fatigue and stress corrosion cracking require much less energy than microvoid coalescence. The highly localised cyclic plasticity characteristic of fatigue is very effective in causing crack growth, especially in an aggressive environment. In stress corrosion the environment enhances fracture in some way. Two common suggestions are dissolution of stressed material, and a reaction to give hydrogen which diffuses into the material and embrittles it, see section 13.4.

FRACTURE	TYPICAL OCCURRENCE	CHARACTERISTICS
• MICROVOID COALESCENCE (DUCTILE FRACTURE)	• STABLE AND UNSTABLE FRACTURE • HIGH STRESS INTENSITY FATIGUE	• IRREGULAR BLUNT CRACK FRONTS • CONSIDERABLE LOCAL PLASTICITY
• CLEAVAGE IN STEELS	• UNSTABLE FRACTURE • INTERMEDIATE AND HIGH STRESS INTENSITY FATIGUE • HYDROGEN EMBRITTLEMENT	• SHARP KINKED CRACKS • BRITTLE FRACTURE
• CONTINUUM FATIGUE (STRIATIONS)	• INTERMEDIATE AND HIGH STRESS INTENSITY FATIGUE	• UNIFORM CRACK FRONTS; LIMITED BLUNTING • MODE I FAVOURED • CONSIDERABLE LOCAL (CYCLIC) PLASTICITY
• SLIP PLANE CRACKING (FACETED FRACTURE)	• FATIGUE CRACK INITIATION AND GROWTH AT LOW STRESS INTENSITIES, ESPECIALLY UNDER TORSIONAL LOADING	• BRANCHED, KINKED AND IRREGULAR CRACK FRONTS; LIMITED BLUNTING • MODE II FAVOURED • LIMITED LOCAL (CYCLIC) PLASTICITY
• CLEAVAGE-LIKE FACETED FRACTURE	• LOW STRESS INTENSITY FATIGUE • STRESS CORROSION	• BRANCHED, KINKED AND IRREGULAR CRACK FRONTS; LIMITED BLUNTING • LIMITED LOCAL PLASTICITY

Figure 13.8. Important kinds of transgranular fracture.

FRACTURE	TYPICAL OCCURRENCE	MECHANICAL CHARACTERISTICS
• MICROVOID COALESCENCE ALONG GRAIN BOUNDARIES	• STABLE AND UNSTABLE FRACTURE AND HIGH STRESS INTENSITY FATIGUE IN HIGH STRENGTH STEELS AND ALUMINIUM ALLOYS	• LIMITED LOCAL PLASTICITY
• LOW DUCTILITY GRAIN BOUNDARY SEPARATION	• SUSTAINED LOAD FRACTURE (INCLUDING CREEP)	• LITTLE EVIDENCE OF PLASTICITY
• BRITTLE GRAIN BOUNDARY SEPARATION	• UNSTABLE FRACTURE IN TEMPER EMBRITTLED STEELS AND REFRACTORY METALS (E.G. TUNGSTEN)	• BRITTLE LOW ENERGY FRACTURE

Figure 13.9. Classification of intergranular fracture.

Intergranular Fracture

Important kinds of intergranular fracture are listed in figure 13.9. The geometric characteristics of intergranular fractures are broadly similar, namely branched and irregular crack fronts with limited blunting. Intergranular fatigue fractures are usually less branched and less irregular than other kinds of intergranular fracture, particularly stress corrosion cracks.

Intergranular fractures require only moderate to low amounts of energy, i.e. they are generally a sign of weakness.

Mixed Fracture Paths

Combinations of different kinds of transgranular fracture and transgranular and intergranular fractures are common. For example, in high strength steels cleavage often occurs as a secondary kind of fracture during fatigue or in combination with transgranular and intergranular microvoid coalescence during unstable fracture.

Since cleavage is a low energy fracture with sharp and fairly uniform crack fronts its occurrence is always detrimental to the fracture mechanics properties. However, the geometric and mechanical characteristics of some fracture paths can oppose each other, e.g.:

(1) Intergranular microvoid coalescence causes a more irregular and branched crack front but the fracture toughness is lower when it occurs in combination with transgranular microvoid coalescence.

(2) Transgranular microvoid coalescence causes irregular and blunt crack fronts but accelerates constant amplitude fatigue crack growth when it occurs with fatigue striations.

The first example is apparently straightforward. Fracture toughness is at least partly determined by stable crack extension, see section 5.2. The limited local plasticity characteristic of intergranular microvoid coalescence outweighs potential increases in fracture toughness owing to branching and irregularity. Note, however, that this rationale is incomplete: it does not explain why intergranular or transgranular fracture paths occur or why they differ in energy. The second example is impossible to explain, even partially, without a detailed consideration of the role of the microstructure.

Both examples will be returned to in the discussion on mechanisms of fracture in section 13.4, under the subheading of fracture by microvoid coalescence.

Effects of Microstructure

There are many microstructural and microstructurally-related features that can play a role in determining the fracture path. The most important are:

- second phases
- particles and precipitates
- grain size
- fibering and texturing owing to mechanical working.

Figure 13.10 illustrates some of these features. Not shown are grain size variations and texturing, which is a preferential orientation of crystal planes. Texturing is important for two phase ($\alpha-\beta$) titanium alloys (α is hexagonal close packed, β is body centred cubic).

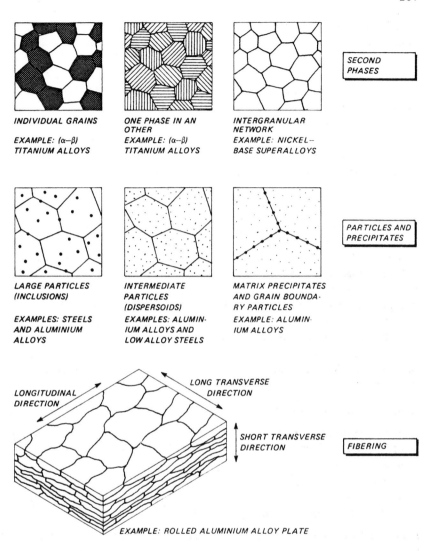

Figure 13.10. Some microstructural and microstructurally-related features in metallic materials.

High strength structural materials usually possess several microstructural features that can influence the fracture path and hence the fracture mechanics properties. The fracture process is therefore often complex, and it is impossible to give an overall, unified description.

Nevertheless, in figure 13.11 we have attempted a survey of microstructural influences on the fracture of high strength steels, aluminium alloys and titanium alloys, which represent three of the four major classes of modern structural materials mentioned in chapter 1. The background to this survey is twofold:

MICROSTRUCTURAL FEATURE	SPECIFIC ASPECTS	EFFECTS ON FRACTURE PATH	EFFECTS ON FRACTURE MECHANICS PROPERTIES
• PHASE MORPHOLOGY IN $\alpha-\beta$ TITANIUM ALLOYS	• CHANGE FROM α AND β GRAINS TO α IN β (SEE FIGURE 13.10)	• CRACK BRANCHING AND IRREGULARITY GREATLY INCREASED	• HIGHER K_{Ic} AND K_{Iscc} • INCREASED RESISTANCE TO FATIGUE CRACK GROWTH
• LARGE PARTICLES IN STEELS AND ALUMINIUM ALLOYS	• RANDOMLY DISTRI-BUTED	• LARGE MICROVOIDS (MAINLY TRANSGRA-NULAR)	• LOWER K_{Ic} • ACCELERATION OF HIGH STRESS INTENSITY FATIGUE
	• INTERGRANULAR CARBIDES IN STEELS	• NUCLEATION OF CLEAVAGE	
• DISPERSOIDS IN STEELS AND ALUMINIUM ALLOYS	• RANDOMLY DISTRI-BUTED	• SHEETS OF SMALL TRANSGRANULAR MICROVOIDS	• LOWER K_{Ic} • ACCELERATION OF HIGH STRESS INTENSITY FATIGUE
	• INCREASED SPACING BETWEEN DISPER-SOIDS IN ALUMINIUM ALLOYS	• FACETED CRACK GROWTH MAINTAINED TO HIGHER ΔK	• TRANSITION FROM REGION I TO REGION II FATIGUE CRACK GROWTH POSTPONED TO HIGHER ΔK
• PRECIPITATES AND PARTICLES IN Al-Zn-Mg-Cu ALLOYS	• LARGER MATRIX PRECIPITATES AND GRAIN BOUNDARY PARTICLES OWING TO OVERAGEING	• INCREASE IN INTER-GRANULAR MICRO-VOID COALESCENCE DURING UNSTABLE FRACTURE	• LOWER K_{Ic} AND K_c
		• NONE FOR STRESS CORROSION	• HIGHER K_{Iscc} AND LOWER REGION II CRACK GROWTH VELOCITY
• LARGER GRAIN SIZE	• OVERAGED Al-Zn-Mg-Cu ALLOYS	• INCREASE IN INTER-GRANULAR MICRO-VOID COALESCENCE	• LOWER K_{Ic} AND K_c
	• STEELS	• FACETED CRACK GROWTH MAIN-TAINED TO HIGHER ΔK	• TRANSITION FROM REGION I TO REGION II FATIGUE CRACK GROWTH POSTPONED TO HIGHER ΔK: HIGHER ΔK_{th}
	• CONVENTIONAL $\alpha-\beta$ TITANIUM ALLOYS	• CRACK BRANCHING AND IRREGULARITY MAINTAINED TO HIGHER ΔK	• INCREASED RESISTANCE TO REGION II FATIGUE CRACK GROWTH
• FIBERING IN ALUMINIUM ALLOYS	• ALIGNMENT OF GRAINS AND PARTICLES	• BRANCHING AND IR-REGULARITY MUCH LESS WHEN CRACK PLANE NORMAL TO SHORT TRANSVERSE DIRECTION (SEE FIGURE 13.10)	• LOWER K_{Ic} AND K_{Iscc} • DECREASED RESISTANCE TO FATIGUE CRACK GROWTH
• CRYSTALLOGRAPHIC TEXTURE IN $\alpha-\beta$ TITANIUM ALLOYS	• PREFERRED ORIEN-TATION OF THE HEXAGONAL α PHASE IN MICROSTRUCTURES WITH INDIVIDUAL α AND β GRAINS (SEE FIGURE 13.10)	• CRACK BLUNTING, BRANCHING AND IRREGULARITY DEPEND ON TEXTURE	• COMPLEX AND STRONG INFLUENCES ON K_{Ic}, K_{Iscc}, ΔK_{th} AND THE RESISTANCE TO FATIGUE CRACK GROWTH

Figure 13.11. Survey of microstructural influences on fracture path and fracture mechanics properties for high strength structural materials.

(1) Relationships between microstructure and fracture properties are better understood for aluminium alloys than for high strength steels and titanium alloys. A review is given in reference 9.

(2) There are basic differences between the classes of materials. Titanium alloys contain very few particles and their fracture properties depend mostly on alloy phase morphology, relative amounts of α and β, and the texture.

Particles are always present in high strength steels and aluminium alloys and have a major effect on fracture toughness, as will be discussed in section 13.4. Also, aluminium alloys are essentially single phase while alloy phases are present only in small amounts in high strength steels, where they are much less important than in titanium alloys.

13.4. Fracture Mechanics and the Mechanisms of Fracture

The mechanisms of fracture in structural materials are incompletely understood. This is particularly true for environmentally influenced crack growth, i.e. environmental fatigue crack propagation and sustained load fracture.

There have been many attempts to model fracture processes using fracture mechanics. The models are necessarily based on continuum behaviour. Their successes −and failures− can provide useful insight into the actual mechanisms of crack growth, and it is this aspect of fracture mechanics that is considered in this final section.

The topics that will be discussed are:

- fracture by microvoid coalescence
- cleavage in steels
- fatigue crack growth
- sustained load fracture
- superposition or competition of sustained load fracture and fatigue.

Fracture by Microvoid Coalescence

Transgranular microvoid coalescence is the typical process by which slow stable tearing and unstable ductile fracture occur. As discussed in section 12.4, the microvoids nucleate at various discontinuities. For steels and aluminium alloys the most important nucleation sites are large particles and dispersoids. In titanium alloys the voids nucleate at the boundaries between α and β phases.

In titanium alloys microvoid nucleation, growth and coalescence to cause fracture is a highly complex process that depends greatly on microstructure and anisotropic plastic deformation owing to texturing. This process is not well understood and will not be discussed further. Interested readers should consult references 10−12 of the bibliography.

A schematic of transgranular microvoid nucleation at particles and subsequent crack extension was shown in figure 12.18 and is given again as figure 13.12. The voids can initiate both at matrix/particle interfaces and as a result of particle fracture. Large particles provide weak spots for the

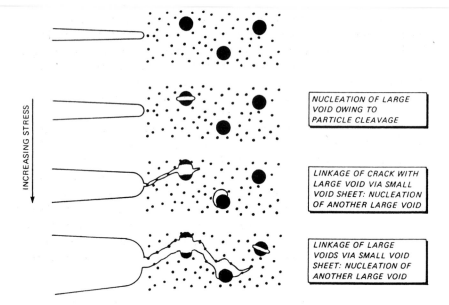

Figure 13.12. Schematic of crack extension by transgranular microvoid coalescence.

nucleation of large voids, which link up to the crack and each other via sheets of small voids nucleated at dispersoids. From this qualitative description two factors are evident:

(1)   Crack tip blunting is important. Blunting will be beneficial to fracture toughness since it lowers the maximum normal stress ahead of the crack, figure 13.3. Therefore void nucleation, growth and coalescence will require higher external stress.

(2)   The sizes and spacings of particles will greatly affect fracture toughness. Larger particles and less distance between them will be detrimental.

These factors are interrelated. Crack tip blunting depends not only on the matrix ductility but also on the ease with which voids nucleate, grow and coalesce with the crack tip: at that moment blunting ceases.

The influence of particle spacing provides an explanation why intergranular microvoid coalescence lowers fracture toughness and why fibering results in lower fracture toughness when the crack plane is normal to the short transverse direction. In both cases the fracture path contains relatively large numbers of particles, i.e. the distance between them is small. However, it is the particle size which determines whether intergranular microvoid coalescence will occur instead of transgranular microvoid coalescence. This is demonstrated by the effect of overageing on unstable fracture in Al-Zn-Mg-Cu alloys, mentioned in figure 13.11.

The contribution of particles to transgranular microvoid coalescence makes it possible to explain why its occurrence accelerates high stress intensity constant amplitude fatigue crack growth. During high stress intensity fatigue microvoid coalescence nucleates at large particles ahead of the crack.

Localized regions of material between the crack tip and the particles also fracture by microvoid coalescence, as in figure 13.12. This process causes rapid local jumps of the crack front such that the overall crack growth rate increases even though the crack front is locally blunter and more irregular.
material class for fracture resistant structures. This is because there is a general trend for fracture toughness to decrease with increasing yield strength,
material class for fracture resistant structures. This is because there is a general trend for fracture toughness to decrease with increasing yield strength,
see figue 13.13. It is most unlikely that any single factor is responsible for this trend. Nevertheless, fractography has shown that the diameters of coalesced microvoids also decrease with increasing yield strength of a class of alesced microvoids also decrease with increasing yield strength of a class of alloys, and has led to the following simple explanation of the general decrease in fracture toughness. Increasing strength raises the attainable stresses in the crack tip region, resulting in earlier nucleation and growth of microvoids during loading and the nucleation and growth of microvoids at many sites that are inactive in a lower strength alloy.

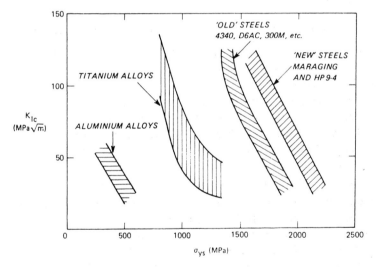

Figure 13.13. General dependency of fracture toughness on yield strength.
After reference 9.

Many attempts have been made to calculate the plane strain fracture toughness, K_{Ic}, from other properties. There are several problems that such modelling faces:

(1) Stress-strain distributions in the plastic zone ahead of the crack must be known or assumed. In this respect strain hardening is very important.

(2) The proper fracture criterion must be chosen. Ductile rupture is strain controlled, i.e. the local strains must exceed a critical value. This is sometimes considered to be the uniform elongation strain

in a tensile test, and in other cases is considered to be the fracture strain at the crack tip.

(3) The critical strain has to be reached or exceeded over a certain distance or volume. A reasonable assumption is that this distance is equal to the particle spacing d. However, there is a complication. The critical strain depends strongly on the stress state, which varies significantly near the crack tip.

(4) Calculation of K_{Ic} is based on the assumption that unstable fracture occurs when the fracture criterion is satisfied. But actual determination of K_{Ic} involves 2% crack extension which, if stable, can cause a significant increase in stress intensity, as mentioned in section 4.8. Consequently, calculated and measured K_{Ic} values need not be comparable.

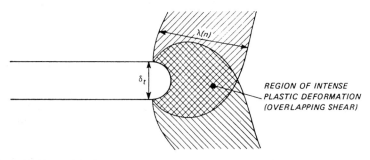

Figure 13.14. Model of plane strain plastic zone at the onset of instability.

In view of these problems it is no surprise that estimates of K_{Ic} are often very inaccurate. One of the more reliable and yet simple models is a semi-empirical one due to Hahn and Rosenfield, reference 13. Figure 13.14 illustrates the basic features of the model. There is a region of intense plastic deformation in the vicinity of the crack tip. The width of this region, λ, depends on the strain hardening characteristics of the material, represented by the strain hardening exponent n. The shear strain at the crack tip is given approximately by

$$\gamma = (\frac{\delta_t}{2})/\lambda(n). \tag{13.2}$$

Two assumptions are now made. First, it is assumed that the average tensile strain, $\bar{\epsilon}$, in the region of intense plastic deformation is approximately $\gamma/2$. Second, the strain distribution is assumed to be linear. Then the maximum tensile strain at the crack tip is

$$\epsilon_{max} = 2\bar{\epsilon} = \gamma = \frac{\delta_t}{2\lambda(n)}. \tag{13.3}$$

At the onset of fracture $\delta_t = \delta_{t\,crit}$, $\epsilon_{max} = \epsilon_f^*$ and $\lambda(n) = \lambda(n)_{crit}$. Thus

$$\epsilon_f^* = \frac{\delta_{t\,crit}}{2\lambda(n)_{crit}}. \tag{13.4}$$

Hahn and Rosenfield measured $\lambda(n)_{crit}$ for a variety of steels and aluminium and titanium alloys. They found that $\lambda(n)_{crit} \sim 0.025\,n^2$ when measured in metres. Also, they argued that the crack tip fracture strain can be related to the true strain, ϵ_f, in a tensile test by $\epsilon_f^* = \epsilon_f/3$. Then

$$\delta_{t\,crit} = \frac{0.05\epsilon_f n^2}{3}. \tag{13.5}$$

Experiments by Robinson and Tetelman (reference 14) have shown that under plane strain conditions the relationship between δ_t and K_I tends to the value predicted by the Dugdale approach (equation 3.16)

$$\delta_t = \frac{K_I^2 (1 - \nu^2)}{\alpha E \sigma_{ys}} \sim \frac{K_I^2}{E \sigma_{ys}}. \tag{13.6}$$

Substituting for $\delta_{t\,crit}$ and K_{Ic} in equation (13.5) gives

$$K_{Ic} = \sqrt{\frac{0.05\epsilon_f n^2 E \sigma_{ys}}{3}} \quad (MPa\sqrt{m}). \tag{13.7}$$

Equation (13.7) was found by Hahn and Rosenfield to be accurate within 30% for eleven different materials.

The derivation of equation (13.7) given here is based on a more recent analysis by Garrett and Knott, reference 15. Alternatively one can arrive at the same result by taking $\epsilon_f^* = \delta_{t\,crit}/\lambda_{t\,crit}$ and $\delta_t \sim K_I^2/2E\sigma_{ys}$, which is the more widely quoted expression for the crack opening displacement under plane strain conditions.

The model of Hahn and Rosenfield contains only macroscopic parameters. The influence of microstructure on fracture toughness is therefore only implicit (i.e. by its effect on these parameters) rather than explicit. An obvious extension of the model is to incorporate the observed behaviour of microvoid nucleation and coalescence. This can be done by specifying that the average strain over the distance d between particles must equal ϵ_f^* for fract-

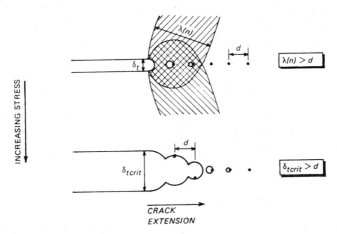

Figure 13.15. Microvoid formation at particles within the region of intense plastic deformation.

ure to occur. Then it appears that in order to obtain even very rough agreement with experimentally determined K_{Ic} values the distance d must be considerably less than $\lambda(n)$ and also less than $\delta_{t_{crit}}$. This is illustrated in figure 13.15.

Several relationships between K_{Ic}, particle spacing and other material properties have been derived. Schwalbe (reference 16) lists two of the more successful relationships as:

$$K_{Ic} = \frac{\sigma_{ys}}{1 - 2\nu} \sqrt{d\pi(1 + n)[\frac{\epsilon_f^* E}{\sigma_{ys}}]^{1+n}} \qquad (13.8)$$

$$K_{Ic} = \sqrt{4.55(\epsilon_f^* + 0.23)dE\sigma_{ys}} \qquad (13.9)$$

These equations illustrate that K_{Ic} depends in a complex way on other material properties. For example, as σ_{ys} increases one may expect ϵ_f^* to decrease, and this decrease is related to a decrease in the effective d (= microvoid diameter) owing to nucleation and growth of microvoids at many more sites. As figure 13.13 shows, the overall result is a trend of decreasing fracture toughness with increasing yield strength for each class of material.

Cleavage in Steels

Cleavage in steels is responsible for the phenomenon of a transition temperature below which brittle, low energy fracture occurs. Because of its great practical importance the occurence of cleavage has been the subject of much experimental and theoretical work. Knott's book on fracture mechanics, reference 17, reviews the micromechanistic theories of cleavage (see also section 12.5) and a model for the dependence of fracture toughness on temperature. This model uses a semi-quantitative description of cleavage initiated at intergranular carbides ahead of the crack tip and will be discussed with the help of figure 13.16. The model proposes that unstable fracture will occur when a critical fracture stress, σ_f, is exceeded by σ_y over a fixed, characteristic distance ahead of the crack tip. The association of cleavage with cracking of intergranular carbides led Knott and coworkers (reference 18) to choose one or two grain diameters as the characteristic distance.

The dependence of fracture toughness on temperature can be explained as follows. At low temperatures the crack is sharp, the material yield stress is high, and little stress intensification is required in order to exceed σ_f over the characteristic distance. Consequently, at failure the plastic zone size is small and K_{Ic} is low. At higher temperatures the crack tip blunts, the material yield stress decreases, and more stress intensification is required for failure, resulting in larger plastic zones and higher K_{Ic} values.

Note that the stress intensification at higher temperatures can exceed $3\sigma_{ys}$. This is because real materials are not elastic-perfectly plastic. Instead crack tip blunting results in strain hardening, which can raise σ_y as high as 4–5 times the nominal yield stress, σ_{ys}. In other words, for real materials the plastic constraint factor, C, (discussed in section 3.5) can be greater than 3.

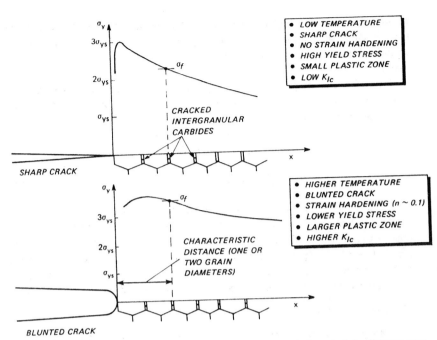

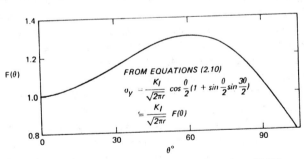

Figure 13.16. Schematic of the critical fracture stress, σ_f, and characteristic distance over which it must be exceeded for unstable cleavage fracture to occur.

The model shown in figure 13.16 appears to be broadly correct. There are, however, two additional and important points:

(1) Cleavage crack nucleation need not occur directly ahead of the crack, i.e. for $\theta = 0°$. This is because the highest values of σ_y occur at $\theta = 50-70°$, see figure 13.17.

Figure 13.17. Variation of σ_y with θ according to equations (2.10).

In fact, there is evidence that so-called 'out-of-plane' cleavage does occur ahead of the main crack, reference 19.

(2) The characteristic distance over which σ_f must be exceeded by σ_y is not simply related to grain size (i.e. one or two grain diameters).

Rather, the characteristic distance is a statistically based quantity depending on the volume of material that must be sampled in order to find a cracked carbide greater than the critical size for cleavage nucleation.

Based on this model Curry (reference 20) has recently given a general expression for the fracture toughness of steels when cleavage occurs:

$$K_{Ic} = \sqrt{\left(\frac{\sigma_f^{n+1}}{\sigma_{ys}^{n-1}}\right)\left(\frac{X}{\beta^{n+1}}\right)} \qquad (13.10)$$

where X is the characteristic distance, which must be determined empirically, and β is a factor between 3 and 5 that expresses the maximum amount of stress intensification near the crack tip in the actual material, i.e. β includes the effects of strain hardening and crack blunting.

Equation (13.10) can be used to predict the cleavage-controlled fracture toughness of a steel at different temperatures and with different microstructures provided that

- the dependences of σ_f and X on microstructure (principally grain size) are known
- the dependences of σ_{ys} and n on temperature and microstructure are known
- the value of β can be estimated from elastic-plastic analysis of crack tip stresses
- some K_{Ic} data are available as empirical checks on σ_f and X.

These requirements restrict the predictive usefulness of equation (13.10) to steels for which a substantial data bank already exists, for example mild steel.

Fatigue Crack Growth

In the previous chapter the mechanisms of fatigue crack initiation and propagation were discussed in some detail and the concept of dislocation movement to cause slip was introduced. This micromechanistic background is necessary to understanding some of the attempts to model fatigue crack growth in fracture mechanics terms.

There are three aspects of fatigue crack growth that will be considered here:

(1) The fatigue crack propagation threshold, ΔK_{th}.
(2) Relations between cyclic plastic zone size, microstructure and regions I and II fatigue crack growth.
(3) Prediction of continuum mode region II fatigue crack growth.

(1) The fatigue crack propagation threshold, ΔK_{th}, depends markedly on
- the elastic modulus
- the microstructure, especially grain size in steels, and phase morphology and texture in titanium alloys
- the environment
- the stress ratio, R.

This list is discouraging. Nevertheless, some simplified models of the threshold condition have achieved remarkably good agreement between prediction and experiment. The models consider either critical fracture stresses and strains or critical shear stresses for slip as the criteria for crack extension. In view of the tendency for near-threshold fatigue crack growth to take place by a combination of modes I and II it seems reasonable that either type of crack extension criterion could apply. We shall here discuss one model of each type and then examine the usefulness of such models in understanding the threshold condition.

Yu and Yan (reference 21) suggested that at threshold the cyclic plastic strain, ϵ_p^c, at a crack tip of finite radius, ρ, becomes equal to the true fracture strain, ϵ_f. They assumed that the reversed (i.e. cyclic) plastic zone is circular and its diameter, r_y^c, can be obtained from the first order estimate of the monotonic plane stress plastic zone by substituting ΔK for K_I and $2\sigma_{ys}$ (reversed plastic flow) for σ_{ys} in equation (3.2), i.e.

monotonic plane stress
$$r_y = \frac{1}{2\pi}\left(\frac{K_I}{\sigma_{ys}}\right)^2 \tag{13.11}$$

cyclic plane strain
$$r_y^c = \frac{1}{2\pi}\left(\frac{\Delta K}{2\sigma_{ys}}\right)^2 \tag{13.12}$$

The plastic strain distribution within the cyclic plastic zone was taken to be

$$\epsilon_p^c = 2\epsilon_{ys}\left(\frac{r_y^c}{r+\rho}\right)^{\frac{1}{1+n}} \tag{13.13}$$

where r is the distance from the crack tip, n is the strain hardening exponent and $2\epsilon_{ys}$ represents reversed plastic flow. Substitution of equation (13.12) into equation (13.13) gives

$$\epsilon_p^c = 2\epsilon_{ys}\left[\frac{\Delta K^2}{8\pi\sigma_{ys}^2(r+\rho)}\right]^{\frac{1}{1+n}} \tag{13.14}$$

At threshold $\Delta K = \Delta K_{th}$ and $\epsilon_p^c = \epsilon_f$ at the crack tip (r = 0). Thus

$$\epsilon_f = 2\epsilon_{ys}\left[\frac{\Delta K_{th}^2}{8\pi\sigma_{ys}^2\,\rho}\right]^{\frac{1}{1+n}} \tag{13.15}$$

or

$$\Delta K_{th} = 2\sigma_{ys}\left(\frac{\epsilon_f}{2\epsilon_{ys}}\right)^{\frac{1+n}{2}}\sqrt{2\pi\rho}. \tag{13.16}$$

Since $\epsilon_{ys} = \sigma_{ys}/E$,

$$\Delta K_{th} = 2\sigma_{ys}\left(\frac{E\epsilon_f}{2\sigma_{ys}}\right)^{\frac{1+n}{2}}\sqrt{2\pi\rho}. \tag{13.17}$$

Yu and Yan then assumed n = 1, which is true only for elastic straining. Then equation (13.17) becomes

$$\Delta K_{th} = E\epsilon_f\sqrt{2\pi\rho}. \tag{13.18}$$

Equation (13.18) is supposed to be valid for R = 0 and a minimum crack tip radius, ρ_{min}, which cannot be less than the Burgers vectors of the dislocations causing plastic deformation. ρ_{min} turns out to be $\sim 0.25-0.29$ nm (nanometres) depending on the material.

Figure 13.18 compares the predictions of equation (13.18) with experimental values of ΔK_{th} for a wide variety of materials. The agreement is very good. This is remarkable in view of the assumption that the cyclic plastic zone is circular. Actually it is nothing of the kind, as figure 13.19 shows.

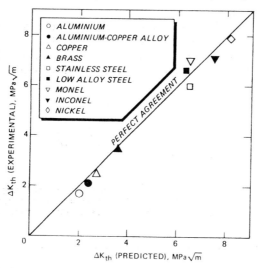

Figure 13.18. Comparison of predicted and experimental values of ΔK_{th} from reference 21.

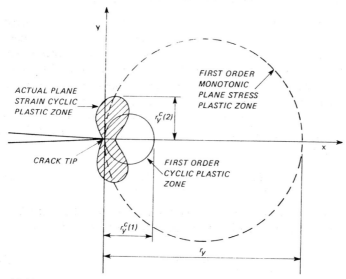

Figure 13.19. Comparison of first order monotonic and cyclic plastic zones with the actual plane strain cyclic plastic zone.

However, the major dimension of the actual cyclic plastic zone is close to the diameter of the first order cyclic plastic zone. Specifically, from reference 22:

$$r_y^c(2) \approx 0.033(\frac{\Delta K}{\sigma_{ys}})^2 = 0.83\, r_y^c(1) \qquad (13.19)$$

Substitution of $r_y^c(2)$ for $r_y^c(1)$ in the analysis of Yu and Yan increases the predicted ΔK_{th} values by $1/\sqrt{0.83} = 1.1$. The agreement is still good, and so it appears that the major dimension of the cyclic plastic zone is more important than its shape.

Note also that Yu and Yan suggested that the crack tip fracture strain is equal to ϵ_f, the true fracture strain in a tensile test, whereas Hahn and Rosenfield argued that for unstable crack extension, equation (13.7), the crack tip fracture strain is $\epsilon_f/3$. This illustrates the uncertainties in the understanding of deformation and fracture at crack tips.

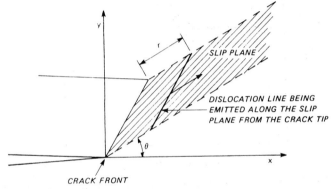

Figure 13.20. Schematic of dislocation emission from a crack tip.

A very different threshold model is the dislocation model of Sadananda and Shahinian, reference 23. A simplified version of the model is shown in figure 13.20. Since fatigue crack growth is usually a consequence of plastic deformation by slip due to the nucleation and movement of dislocations, it is assumed that the threshold condition corresponds to the minimum shear stress required to nucleate and move a dislocation from the crack tip. Sadananda and Shahinian derived a general expression for this shear stress. For a dislocation moving directly away on a slip plane with $\theta = 45°$ the general expression reduces to

$$\tau = \frac{\mu}{4\pi(1-\nu)}(\ln\frac{4r}{b} + \frac{b}{r}) + \frac{\gamma_e}{b\sqrt{2}} + \frac{\sigma_{ys}}{2} \qquad (13.20)$$

where μ is the shear modulus, γ_e is the surface energy and b is the Burgers vector. Now for a sharp crack we can write from equations (2.10):

$$\tau_{xy} = \frac{K_I}{\sqrt{2\pi r}}\sin\frac{\theta}{2}\cos\frac{\theta}{2}\cos\frac{3\theta}{2}. \qquad (13.21)$$

For a dislocation to be nucleated at the crack tip τ_{xy} should be at least equal to τ and r should be at least b. Hence the threshold condition is given by

$$K_{I_{th}} = \frac{\tau\sqrt{2\pi b}}{\sin\frac{\theta}{2}\cos\frac{\theta}{2}\cos\frac{3\theta}{2}} \qquad (13.22)$$

For R = 0 the quantity $K_{I_{th}}$ can be replaced by ΔK_{th}. Also, $\theta = 45°$. Then equation (13.22) reduces to

$$\Delta K_{th} = 18.5\tau\sqrt{b} \qquad (13.23)$$

and for different R values Sadananda and Shahinian suggest

$$\Delta K_{th} = K_{I_{th}}(1 - R). \qquad (13.24)$$

The model gives good agreement between predicted and experimental values of ΔK_{th}, although there is some uncertainty as to the actual values of surface energy, γ_e, for many metals and alloys, and hence the value of τ.

Both of the models just described show that ΔK_{th} depends strongly on the elastic moduli (E or μ), in agreement with many experimental results and with other models. Also, the cyclic plastic zone size is an important parameter. This is emphasized by a model due to Taylor (reference 24) in which the cyclic plastic zone size is set equal to the grain size and thereby allows reasonable prediction of the dependence of ΔK_{th} on grain size in steels.

With insight provided by the models it is possible to explain the existence of fatigue thresholds. Figure 13.21 is a schematic of how fatigue crack growth depends on the relative sizes of the cyclic plastic zone and the grain size. When the cyclic plastic zone is significantly larger than the grain size the high local stress concentrations induce slip on several sets of crystal planes in each grain and also ensure that slip spreads across barriers like grain boundaries. The plastic deformation is therefore homogeneous and the fatigue crack propagates by a continuum mechanism.

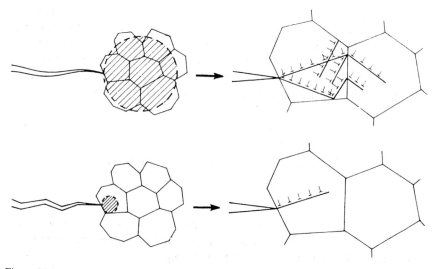

Figure 13.21. Dependence of fatigue crack growth on cyclic plastic zone size and grain size.

However, when the cyclic plastic zone size is less than the grain size the local stress concentrations are sufficient to activate only one or a few slip planes. Microstructural barriers then exert more influence on the spreading of slip to neighbouring grains, and are assisted in this by the geometric effects of faceted modes I and II crack growth, i.e. lower mode I stress intensities at the tips of kinked cracks and increased crack closure, as discussed in section 13.2.

Eventually the cyclic plastic zone size becomes so small compared to the grain size that slip can no longer spread across microstructural barriers. The dislocations pile up on the activated slip plane(s). This increases the resistance to dislocation emission from the crack tip until no more are emitted: the fatigue threshold has been reached. Cyclic slip can still occur, but is confined to to-and-fro movement of dislocations along the slip plane as far as a microstructural barrier (typically a grain boundary, but in aluminium alloys dispersoid particles may also act as barriers, reference 25).

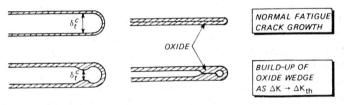

Figure 13.22. Oxide wedge mechanism of raising ΔK_{th}. After reference 26.

The foregoing explanation is mechanical and takes no account of the environment. An aggressive environment usually lowers ΔK_{th}, but sometimes there is little or no change or even an increase in ΔK_{th}. As mentioned in section 9.4, in some instances (mainly for steels) an increase in ΔK_{th} is due to corrosion contributing to crack closure, i.e. the environmental effect is largely mechanical. The way this occurs is illustrated in figure 13.22. In the near threshold region of crack growth there is an enhanced build-up of corrosion product (oxide) on the crack surfaces close behind the crack tip. This oxide build-up acts as a wedge and reduces the cyclic crack tip opening displacement δ_t^c, which under plane strain conditions is approximately given by

$$\delta_t^c = 0.5 \frac{\Delta K^2}{E2\sigma_{ys}^c} \tag{13.25}$$

where σ_{ys}^c is the cyclic yield stress. For many materials $\sigma_{ys}^c \sim \sigma_{ys}$, i.e.

$$\delta_t^c \approx 0.25 \frac{\Delta K^2}{E\sigma_{ys}}. \tag{13.26}$$

Irrespective of the exact value of the cyclic plastic zone size, r_y^c is directly proportional to δ_t^c. Thus the oxide wedge reduces the cyclic plastic zone size and this is reflected in an increase in ΔK_{th}.

(2) The significance of the cyclic plastic zone size is not limited to the threshold condition for fatigue crack growth. Recently it has become evident that there are also relations between cyclic plastic zone size, microstructure and regions I and II fatigue crack growth, depending on the type of material.

Figure 13.23 schematically illustrates transitions in the fatigue crack growth rate curves for steels and titanium and aluminium alloys. These transitions are as follows:

- For steels the region I–region II transition point corresponds to the cyclic plastic zone size equalling the grain size.
- For titanium alloys a knee in the region II crack growth rate curve corresponds to the cyclic plastic zone size equalling the grain size.
- For aluminium alloys there appear to be at least three transition points. At the region I–region II transition point (T_I) the cyclic plastic zone size is approximately equal to the spacing between dispersoid particles. For the region II transition points T_2 and T_3 the cyclic plastic zone size correlates with subgrain size and grain size, respectively. (Subgrains are regions within a grain that differ slightly in crystal orientation with respect to each other. The boundaries between subgrains are dislocation arrays that are barriers to slip, though less effective than grain boundaries.)

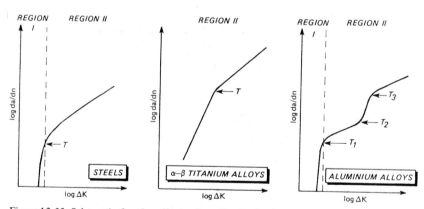

Figure 13.23. Schematic for the effects of cyclic plastic zone size on the fatigue crack growth rate curve: T = transitions influenced by the relation between cyclic plastic zone size and microstructure (see text). After references 25, 27 and 28.

The transition points in steels and conventional titanium alloys represent a change from faceted fracture at lower ΔK to a continuum mechanism of crack propagation. This is as expected from the correlation between the cyclic plastic zone size and the grain size at these transitions. However, for aluminium alloys the transition point T_3 does not correspond to a change from faceted to continuum-type crack growth. Instead, this change occurs at T_2, reference 29. A possible explanation is that even though the cyclic plastic zone is smaller than the grain size below T_3, the misorientations of the subgrains cause slip to be fairly homogeneous. Only when the cyclic plastic zone is smaller than the subgrain size, below T_2, is slip confined to one or a few sets of crystal planes.

(3) Finally, we shall consider briefly the prediction of continuum mode region II fatigue crack growth. Many empirical models have been formulated (see section 9.2) but they do not contribute to understanding the crack growth process. Only a few attempts have been made at absolute predictions of fatigue crack growth rates. The problems to be faced are similar to those for modelling fracture toughness: stress-strain distributions in the plastic zone

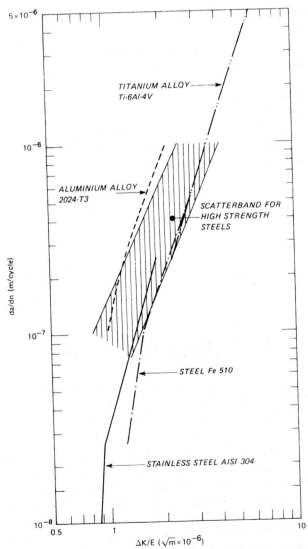

Figure 13.24. Fatigue crack growth rates as functions of $\Delta K/E$ for the data of figu'

ahead of the crack must be known or assumed, and the proper criteria for crack advance must be chosen. In this respect all models are severely limited, since they consider purely mechanical crack extension, i.e. environmental effects cannot be accounted for. This point is often overlooked when model predictions are compared with experimental data. Nearly always the data are for fatigue tests in air, which has a strong environmental influence compared to vacuum or dry gases.

The models fall into two main categories based on the criteria for crack advance, which is assumed to occur during each load cycle by either

- Alternating shear at the crack tip, i.e. incompletely or totally irreversible slip on alternate sets of crystal planes.
- Exceedance of the true fracture strain over a certain distance.

Alternating shear models predict relations of the form

$$\frac{da}{dn} = C \frac{\Delta K^2}{E\sigma_{ys}},$$

(13.27)

i.e. linear log-log behaviour with a slope m = 2. In fact m is rarely 2, nor is m always constant during region II fatigue crack growing. Note that equation (13.27) predicts that da/dn is inversely proportional to E. This relation is supported by experimental data. In figure 13.24 the crack growth rate data of figure 9.5 are plotted as functions of $\Delta K/E$ instead of ΔK. The widely varying crack growth rate curves for various materials are brought together into a single (wide) scatterband.

The simplest model is that of Pelloux, reference 30. In this model all the slip takes place at the crack tip and contributes to crack advance, as shown in figure 13.25.

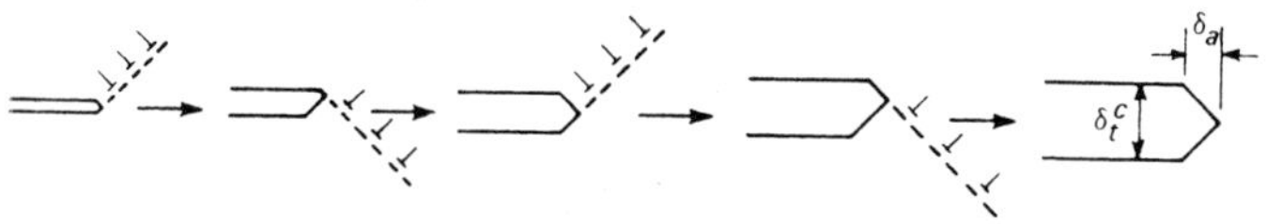

Figure 13.25. Alternating shear model of fatigue crack growth.

It is evident that the crack advance, δ_a, is one-half the cyclic crack tip opening displacement, δ_t^c. Thus from equation (13.25) crack advance under plane strain conditions is given by

$$\delta_a = \frac{\delta_t^c}{2} = \frac{\Delta K^2}{8E\sigma_{ys}^c},$$

(13.28)

where δ_a is the crack advance per cycle, i.e. δ_a = da/dn.

Other alternating shear models allow only part of the slip to contribute to crack advance: the remainder causes crack blunting. Kuo and Liu (reference 31) proposed a model based on both theoretical considerations and experimental determinations of COD and strains near the crack tip. They obtained the following semi-empirical formula:

$$\frac{da}{dn} = \frac{0.019(1 - \nu^2)\Delta K^2}{E\sigma_{ys}^c}. \qquad (13.29)$$

Figure 13.26 compares the models of Pelloux and Kuo and Liu with data for two titanium alloys tested *in vacuo*, i.e. purely mechanical crack extension. The model of Kuo and Liu is clearly in better agreement with the test data. This shows that slip in the crack tip region mainly causes crack blunting. Only a small amount of slip contributes to actual crack advance.

Of the models using a crack advance criterion based on true fracture strain, that of Antolovich et al. (reference 32) provides the most insight into the mechanisms of region II fatigue crack growth. The model will be discussed

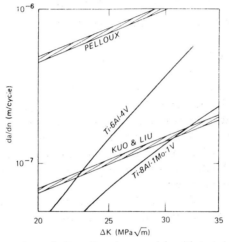

Figure 13.26. Comparison of alternating shear models with test data.

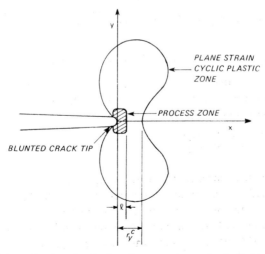

Figure 13.27. Schematic of cyclic plastic and process zones ahead of a propagating fatigue crack.

using figure 13.27, which shows that the fatigue crack is assumed to be blunt and that within the cyclic plastic zone there is a 'process zone'. Note also that r_y^c is defined as the plane strain cyclic plastic zone size along the x axis.

Antolovich et al. derived a rather complicated expression for the fatigue crack growth rate:

$$\frac{da}{dn} = 4\left(\frac{0.7\alpha}{E\sigma_{ys}^{1+S}\epsilon_f}\right)^{\frac{1}{\beta}}\left(\frac{1}{\varrho^{1/\beta}-1}\right)(\Delta K)^{\frac{2+S}{\beta}},$$ (13.31)

where S and α are defined by

$$r_y^c = \alpha\left(\frac{\Delta K}{\sigma_{ys}}\right)^{2+S}$$ (13.32)

and β is the Coffin-Manson low cycle fatigue exponent, which usually lies between 0.4 and 0.6. Equation (13.32) shows that r_y^c is not taken to be proportional to $(\Delta K/\sigma_{ys})^2$. This is because crack tips in real materials are blunt to some extent. The parameter S accounts for crack blunting, and in general is a small number $\sim \pm0.1-0.2$.

For a sharp crack S = 0, and the value of r_y^c can be obtained directly from equation (3.22) in chapter 3 by substituting $2\sigma_{ys}$ (reversed plastic flow) for σ_{ys}. The result is

$$r_y^c = \frac{1}{24\pi}\left(\frac{\Delta K}{\sigma_{ys}}\right)^2$$ (13.33)

Equation (13.31) is the result of modelling crack growth in the process zone in terms of low cycle fatigue. If S and β are known the equation can be fitted to actual data in order to determine the process zone size, ϱ. This acts as a check on the model: there should be a reasonable physical interpretation of ϱ to support the assumption that continuum mode region II fatigue crack growth is the result of a low cycle fatigue process occurring near the crack tip. In fact, ϱ is about the size of dislocation cells, which are polygonal arrangements of dense tangles of dislocations that form during both low cycle fatigue and region II fatigue crack growth. The assumptions of the model therefore appear to be justified (for a more detailed treatment see reference 33).

Sustained Load Fracture

A comprehensive treatment of mechanisms of sustained load fracture is a forbidding task well beyond the scope of this course. The discussion will be restricted to a few examples where fracture mechanics concepts have been used to study such mechanisms. These examples are:

(1)   Time-to-failure (TTF) tests in modes I and III.
(2)   Cleavage during stress corrosion of titanium alloys.
(3)   Susceptibility of steels to hydrogen embrittlement.
(4)   Analysis and modelling of crack growth rates.

(1) The background to TTF testing in modes I and III is that sustained load fracture for some material-environment combinations may occur by hydrogen embrittlement as a result of diffusion of hydrogen to a location ahead

of the crack. The increased hydrogen concentration at this location then results in cracking that links up with the main crack.

In such cases any factor that promotes hydrogen diffusion to a location ahead of the main crack should increase the susceptibility to sustained load fracture. One possibility is dilatation (three-dimensional expansion) of the crystal lattice owing to a state of hydrostatic stress, i.e. a state in which, strictly speaking, all three principal stresses are equal. In the elastic stress fields of cracks the three principal stresses are not all equal to each other, even in plane strain. However, it is convenient to split the triaxial stress state into a hydrostatic component

$$\sigma_h = \tfrac{1}{3}(\sigma_1 + \sigma_2 + \sigma_3) \qquad (13.34.a)$$

and a stress deviator s

$$s_1 = \sigma_1 - \sigma_h, \quad s_2 = \sigma_2 - \sigma_h, \quad s_3 = \sigma_3 - \sigma_h. \qquad (13.34.b)$$

From equations (2.13) and (2.15) in chapter 2 it follows that

$$\text{mode I plane strain} \qquad \sigma_h = \frac{2(1 + \nu)K_I}{3\sqrt{2\pi r}} \cos\frac{\theta}{2} \qquad (13.35)$$

$$\text{mode III} \qquad \sigma_h = 0. \qquad (13.36)$$

In other words, for mode I loading σ_h reaches a maximum directly ahead of the crack, but for mode III loading there is no hydrostatic stress component.

Thus if stress-assisted hydrogen diffusion and embrittlement play a role in sustained load fracture, then comparison tests in mode I and mode III should show differences in susceptibility. This does, in fact, appear to be the case for stress corrosion cracking of steels and aluminium and titanium alloys. An example is given in figure 13.28, which shows that the susceptibility of a titanium alloy to stress corrosion cracking was considerable in mode I but negligible in mode III.

It should be noted that there are also hydrogen embrittlement models for which the concentration of hydrogen in the crystal lattice does not depend on a hydrostatic stress component, e.g. transport of hydrogen by dislocations

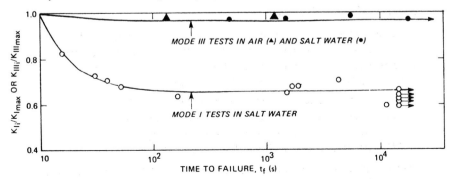

Figure 13.28. Stress corrosion susceptibility of precracked titanium alloy Ti-8Al-1Mo-1V specimens under mode I and mode III loading. After reference 34.

or the hindering of dislocation movement by hydrogen. In such cases there need not be any difference in susceptibility under mode I and mode III loading.

(2) The concept of stress-assisted hydrogen diffusion and embrittlement has also been examined with respect to α phase cleavage during stress corrosion of titanium alloys, reference 35. In highly textured alloys a mode I crack can be approximately parallel to the cleavage planes in most grains. Under these conditions the stress corrosion fracture path is very flat, i.e. cleavage occurs directly ahead of the crack. Figure 13.29 shows that this is evidence for hydrogen embrittlement control of the cleavage location rather than mechanical control, since σ_h is a maximum directly ahead of the crack ($\theta = 0°$) but σ_y is a maximum at $\theta = 65°$. Mechanical control would be expected to cause out-of-plane cleavage, discussed earlier in this section with respect to steels (see the discussion to figure 13.17).

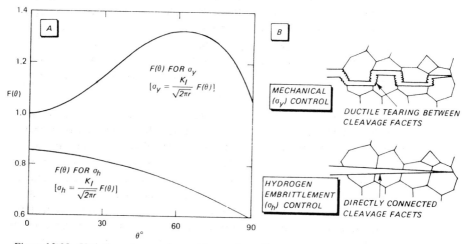

Figure 13.29. Variation of σ_y and σ_h with θ and predicted stress corrosion fracture paths for a mode I crack parallel to the cleavage planes in a highly textured titanium alloy.

(3) Apart from the more generally applicable TTF comparison tests in modes I and III, fracture mechanics analyses of the susceptibility to sustained load fracture are concerned mostly with hydrogen embrittlement of steels, an important practical problem. For example, steel pipelines and pressure vessels are used for transport and storage of hydrogen-containing environments, and power generating equipment may have to operate in a hydrogen atmosphere. These are examples where *external* hydrogen may be a problem. *Internal* hydrogen, i.e. residual hydrogen within the metal, is also important, especially for welds in high strength structural steel, and ultrahigh strength steels as used in most aircraft landing gear and many other components where high strength is essential.

With respect to external hydrogen embrittlement, fracture mechanics stress field solutions have been used to distinguish between two types of theory:

- The adsorption theory, which assumes that hydrogen is adsorbed at the crack tip and changes the surface energy of the crystal lattice, thereby lowering the fracture resistance. The surface energy change, and hence the amount of ambrittlement, depend on the external gas pressure.

- The absorption + decohesion theory. This is another name for stress-assisted hydrogen diffusion ahead of the crack tip and the reaching of a sufficient concentration of hydrogen to cause cracking that links up with the main crack. The theory is applicable not only to hydrogen embrittlement per se, but also to stress corrosion of steels (discussed later in this section). Again, the external gas pressure determines the amount of embrittlement: higher pressure decreases K_{Ith}.

Details of the very complex analyses of external hydrogen embrittlement will not be given here: reference 36 of the bibliography contains a comprehensive review. However, the results are important and appropriate:

- the adsorption theory fails to predict realistic K_{Ith} values
- absorption + decohesion theories which require lattice decohesion to occur very close to the crack tip provide the best predictions of K_{Ith} as a function of external gas pressure.

The physical significance of the second result is that the experimental data with which the models were compared must have been obtained from materials with sharp cracks, such that the maximum value of σ_h (and hence σ_x, σ_y and σ_z) was well within the plastic zone, compare figure 13.3. There is thus a clear correlation between crack tip sharpness and the susceptibility to hydrogen embrittlement. This agrees with the fact that steels with higher yield strengths and less capacity for plastic deformation are more susceptible.

Internal hydrogen embrittlement seems at first to be a different phenomenon from external hydrogen embrittlement: after all, there is no apparent source of gaseous hydrogen. However, van Leeuwen (reference 37) has provided a model to show that the absorption + decohesion theory can be applied

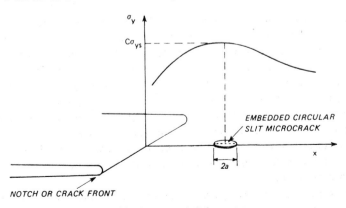

Figure 13.30. Model for internal hydrogen embrittlement.

to internal as well as external hydrogen embrittlement. The model is illustrated in figure 13.30. Ahead of a notch or crack a microcrack forms by decohesion at the location of maximum hydrogen concentration. This is also the location of maximum hydrostatic stress and the maximum value of σ_y, $C\sigma_{ys}$ (C is the plastic constraint factor, previously mentioned in this section with respect to cleavage in steels).

Formation of the microcrack results in a local decrease in hydrostatic stress. Hydrogen can no longer be held in the lattice at the same concentration, and therefore enters the microcrack as gas, building up to an equilibrium pressure, P_e. The stress intensity factor for the microcrack can be obtained from the well known solution in section 2.8, i.e.

$$K_I = \frac{2}{\pi}\sigma\sqrt{\pi a} = \frac{2}{\pi}(C\sigma_{ys} + P_e)\sqrt{\pi a}. \qquad (13.37)$$

Equation (13.37) shows that an increase in P_e will increase K_I. At the same time, by analogy with external hydrogen embrittlement, an increase in P_e should decrease K_{Ith} for the microcrack. Thus critical combinations of K_I and P_e will result in exceedance of K_{Ith} for the microcrack, which then will propagate back to the main crack or notch.

When the microcrack links up with the main crack or notch the gas pressure drops to zero. Thus crack growth stops until another microcrack is nucleated ahead of the new crack front. This means that the total crack propagation process must take place in discrete steps, and this is exactly what happens.

(4) The last topic to be discussed is the analysis and modelling of sustained load crack growth rates. As in the case of fatigue crack growth, of the many models proposed for sustained load crack growth only a few are quantitative in fracture mechanics terms. The difficulties of analysis are formidable, and the phenomenological behaviour is often complex. Models predicting a dependence of da/dt on K_I include:

- stress-assisted dissolution of the crack tip material (region I crack growth)
- stress-assisted diffusion of hydrogen (absorption + decohesion) for regions I and II crack growth in steels owing to stress corrosion and external and internal hydrogen embrittlement
- the capillary model of stress corrosion.

The stress-assisted dissolution model predicts a linear dependence of $\log da/dt$ on K_I, i.e. region I crack growth. The type of equation obtained is

$$\log\frac{da}{dt} = \log A + \frac{2.3}{RT}\left(\frac{2V^*K_I}{\sqrt{\pi\rho}} - E^*\right) \qquad (13.38)$$

where A is a constant, R is the gas constant, T is temperature, ρ is the crack tip radius, E^* is the activation energy of dissolution at zero load, and V^* is an 'activation volume', whose precise physical meaning is unknown. A more serious objection to the model is that it requires extremely high stresses $\sim E/20$ to exist in the crack tip region, reference 38. This means that the model cannot be applied to metallic materials, but it may be appropriate to ceramics.

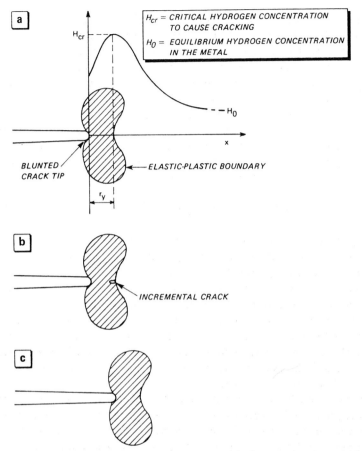

Figure 13.31. Hydrogen absorption + decohesion model of stress corrosion cracking in high strength steels.

A fairly successful model for sustained load crack growth due to stress-assisted diffusion of hydrogen has been derived by van Leeuwen (reference 39) for stress corrosion of high strength steels. Van Leeuwen proposed that the concentration of hydrogen reaches a maximum at the elastic-plastic boundary directly ahead of the crack. When the maximum hydrogen concentration reaches a critical value, H_{cr}, decohesion occurs at the elastic-plastic boundary to cause cracking that links up with the main crack, as shown schematically in figure 13.31. Overall crack propagation is treated as a series of such steps, as in the case of internal hydrogen embrittlement.

Note that it is an assumption of the model that the crack tip is sufficiently blunt for the hydrostatic stress to be a maximum at the elastic-plastic boundary directly ahead of the crack. This is a significant difference from what were found to be the requirements for predicting $K_{I_{th}}$ in gaseous hydrogen, i.e. sharp crack tips and lattice decohesion commencing well within the plastic zone. However, there is ample evidence that a stress corrosion crack can be blunted not only by plasticity but also by corrosion.

To account for crack tip blunting van Leeuwen used the mode I elastic stress field equations derived by Creager and Paris, i.e. equations (13.1) discussed at the beginning of this chapter. The following general expression for the stress intensity factor K_I was obtained, assuming $\theta = 0$ in equation (13.1):

$$K_I = \sigma_y \sqrt{2\pi r} \left(\frac{2r}{2r + \rho} \right). \tag{13.39}$$

Substituting $C\sigma_{ys}$ for σ_y and $r = r_y + \rho/2$ gives

for a blunt crack $\quad K_I = \dfrac{C\sigma_{ys}\sqrt{2\pi}}{r_y + \rho} \left(r_y + \dfrac{\rho}{2} \right)^{\frac{3}{2}}$

$\hspace{12cm} (13.40)$

for a sharp crack $\quad K_I = C\sigma_{ys}\sqrt{2\pi r_y}.$

where r_y is the distance along the x-axis to the elastic plastic boundary and is also the increment of crack growth. A problem is that equation (13.40) will have to be solved by iteration. The actual value of r_y is not known since it will depend on the plastic constraint factor C (compare section 3.5). But C is not known and can only be estimated by substituting $r = r_y + \rho/2$ in equations (13.1) and calculating the principal stresses σ_1 and σ_2 for $\theta = 0°$. Substituting these stresses in the von Mises yield criterion ($\sigma_3 = \nu(\sigma_1 + \sigma_2)$) results in

$$C = \frac{2(r_y + \rho)}{\sqrt{3\rho^2 + (2r_y + \rho)^2 (1 - 2\nu)^2}}. \tag{13.41}$$

Thus C in its turn depends on what value of r_y is substituted.

Van Leeuwen determined a very complicated expression for the relation between da/dt and the hydrogen concentration. Since the critical hydrogen concentration, H_{cr}, depends on r_y and ρ, da/dt will depend on r_y and ρ as well. Choosing various values of r_y and ρ, solving equations (13.40) and (13.41) iteratively, and calculating da/dt from the values of r_y and ρ resulted in a series of curves relating da/dt and K_I. An example is given in figure 13.32. Good fits to experimental data are possible, especially for region II crack growth.

Note that the choice of ρ is always somewhat arbitrary. However, the value of r_y which is used in calculating K_I also determines da/dt, i.e. r_y is not a second arbitrary parameter for data fitting.

Van Leeuwen also showed that the da/dt–K_I curve can be estimated remarkably well using equations (13.40) and the simple expression

$$\frac{da}{dt} = \frac{4Dr_y}{(r_y + \rho/2)^2}. \tag{13.42}$$

where D is the diffusion constant for hydrogen in steel ($\sim 2 \times 10^{-7} cm^2/s$ at room temperature).

The capillary model of stress corrosion (reference 40) proposes that under certain conditions a crack may be incompletely penetrated by liquid, so that if crack propagation occurs the liquid must flow down the crack to maintain

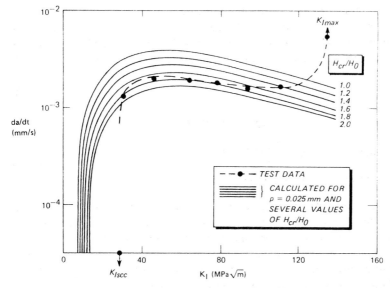

Figure 13.32. Calculated and measured stress corrosion crack growth rates for a high strength steel in salt water. After reference 39.

the propagation. If this dynamic capillary fluid flow is rate controlling the model predicts the existence of regions I and II crack growth. The model also predicts dependences of the crack growth rate curve on fluid viscosity and external gas pressure. Limited testing of the model has been done. The most that can be said is that capillary flow is likely to be a factor in stress corrosion cracking, but it is not known whether it can be rate controlling. It is worth noting, however, that capillary flow could be rate controlling for crack growth due to liquid metal embrittlement, since the velocities of cracking are often extremely high, $\sim 1-10$ cm/s.

Superposition or Competition of Sustained Load Fracture and Fatigue

As mentioned in section 9.4, fatigue crack growth in aggressive environments can be enhanced by sustained load fracture during each load cycle. There are two fairly simple models which try to account for this effect:

- the superposition model, reference 41
- the process competition model, reference 42.

The superposition model proposes that the overall crack growth rate is the sum of a baseline fatigue component and a component due to sustained load fracture. On the other hand, the process competition model assumes that fatigue and sustained load fracture are mutually competitive and that the crack will grow at the fastest available rate, whether that is the baseline fatigue crack growth rate or the crack growth per cycle owing to sustained load fracture.

The models can be expressed formally in the following way:

superposition model $\qquad (\frac{da}{dn})_{tot} = (\frac{da}{dn})_B + \int \frac{da}{dt} \cdot K_I(t)dt \qquad (13.43)$

process
competition $\qquad (\frac{da}{dn})_{tot} = (\frac{da}{dn})_B \qquad$ for $(\frac{da}{dn})_B > \int \frac{da}{dt} \cdot K_I(t)dt$
model

$\qquad = \int \frac{da}{dt} \cdot K_I(t)dt \ \ \text{for} \ (\frac{da}{dn})_B < \int \frac{da}{dt} \cdot K_I(t)dt$

$\qquad (13.44)$

where the integral in equations (13.43) and (13.44) is taken over one cycle of the fatigue loading and incorporates the effects of frequency, f, and stress ratio, R, via $K_I(t)$.

Figure 13.33 is a schematic of the way in which the superposition model can be used to predict the overall crack growth rate curve for constant amplitude sinusoidal loading. For each ΔK value of interest the integral in equation (13.43) can be obtained as the area A under the curve relating da/dt and the

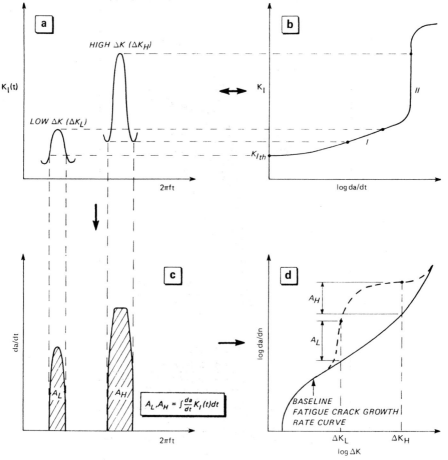

Figure 13.33. Schematic of the superposition of sustained load fracture on fatigue in order to construct the overall crack growth rate curve.

load cycle waveform and frequency characteristics (= $2\pi ft$ for sinusoidal loading). Addition of each A value to the baseline fatigue crack growth rate at the same ΔK results in the predicted overall crack growth rate curve.

A similar procedure can be used for the process competition model, except that the predicted overall crack growth rate curve is given either by the baseline fatigue crack growth rate or by the integral in equations (13.44).

A comparison of the predictions of both models with experimental data is given in figure 13.34. The process competition model gives slightly better predictions, but both models significantly overestimate the overall crack growth rates at each frequency. This is a general trend when the environment is liquid rather than gaseous, i.e. when stress corrosion cracking occurs during each load cycle. The reason is that cyclic closing and opening of the crack provide a pumping action that mixes the liquid near the crack tip, and when the crack is open it takes a finite time for the liquid to re-establish the conditions necessary for stress corrosion.

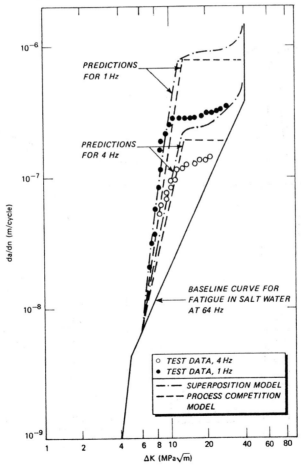

Figure 13.34. Comparison of model predictions with test data for 835M30 steel undergoing stress corrosion during fatigue in salt water at R = 0.5.

There is no general trend of over- or under-estimation when the models are used to predict overall crack growth rates in gaseous environments, and agreement with test data can be good or poor. Thus, in summary it seems fair to state that both models give an indication of the effect of cycle frequency on the overall crack growth rates, but neither is sufficiently accurate for determining whether superposition or competition of sustained load fracture and fatigue actually occur.

13.5. Bibliography

1. Creager, M. and Paris, P.C., *Elastic Field Equations for Blunt Cracks with Reference to Stress Corrosion Cracking*, International Journal of Fracture Mechanics, Vol. 3, pp. 247–252 (1967).
2. McGowan, J.J. and Smith, C.W., *A Plane Strain Analysis of the Blunted Crack Tip Using Small Strain Deformation Plasticity Theory*, Advances in Engineering Science, NASA, Vol. 2, pp. 585–594 (1976): Hampton, Virginia.
3. Gerberich, W.W. and Jatavallabhula, K., *Quantitative Fractography and Dislocation Interpretations of the Cyclic Cleavage Crack Growth Process*, Acta Metallurgica, Vol. 31, pp. 241–255 (1983).
4. Dorward, R.C. and Hasse, K.R., *Flaw Growth of 7075, 7475, 7050 and 7049 Aluminium Plate in Stress Corrosion Environments*, Final Report Contract NAS8–30890, Kaiser Aluminium and Chemical Corporation Centre for Technology (1976): Pleasanton, California.
5. Schijve, J., *The Effect of an Irregular Crack Front on Fatigue Crack Growth*, Engineering Fracture Mechanics, Vol. 14, pp. 467–475 (1981).
6. Forsyth, P.J.E. and Bowen, A.W., *The Relationship between Fatigue Crack Behaviour and Microstructure in 7178 Aluminium Alloy*, International Journal of Fatigue, Vol. 3, pp. 17–25 (1981).
7. Minakawa, K. and McEvily, A.J., *On Crack Closure in the Near-Threshold Region*, Scripta Metallurgica, Vol. 15, pp. 633–636 (1981).
8. Suresh, S., *Crack Growth Retardation due to Micro-Roughness: a Mechanism for Overload Effects in Fatigue*, Scripta Metallurgica, Vol. 16, pp. 995–999 (1982).
9. Wanhill, R.J.H., *Microstructural Influences on Fatigue and Fracture Resistance in High Strength Structural Materials*, Engineering Fracture Mechanics, Vol. 10, pp. 337–357 (1978).
10. Greenfield, M.A. and Margolin, H., *The Mechanism of Void Formation, Void Growth and Tensile Fracture in an Alloy Consisting of Two Ductile Phases*, Metallurgical Transactions, Vol. 3, pp. 2649–2659 (1972).
11. Bowen, A.W., *The Influence of Crystallographic Orientation on the Fracture Toughness of Strongly Textured Ti-6Al-4V*, Acta Metallurgica, Vol. 26, pp. 1423–1433 (1978).
12. Tchorzewski, R.M. and Hutchinson, W.B., *Anisotropy of Fracture Toughness in Textured Titanium-6Al-4V Alloy*, Metallurgical Transactions A, Vol. 9A, pp. 1113–1124 (1978).
13. Hahn, G.T. and Rosenfield, A.R., *Sources of Fracture Toughness: the Relation between K_{Ic} and the Ordinary Tensile Properties of Metals*, Applications Related Phenomena in Titanium Alloys, ASTM STP 432, pp. 5–32 (1968): Philadelphia.
14. Robinson, J.N. and Tetelman, A.S., *The Determination of K_{Ic} Values from Measurements of the Critical Crack Tip Opening Displacement at Fracture Initiation*, Paper II–421, Third International Conference on Fracture, Munich (1973).
15. Garrett, G.G. and Knott, J.F., *The Influence of Compositional and Microstructural Variations on the Mechanism of Static Fracture in Aluminium Alloys*, Metallurgical Transactions A, Vol. 9A, pp. 1187–1201 (1978).
16. Schwalbe, K.-H., *On the Influence of Microstructure on Crack Propagation Mech-

anisms and Fracture Toughness of Metallic Materials, Engineering Fracture Mechanics, Vol. 9, pp. 795–832 (1977).

17. Knott, J.F., *Fundamentals of Fracture Mechanics*, Butterworths (1973); London.

18. Ritchie, R.O., Knott, J.F. and Rice, J.R., *On the Relationship between Critical Tensile Stress and Fracture Toughness in Mild Steel*, Journal of Mechanics and Physics of Solids, Vol. 21, pp. 395–410 (1973).

19. Stonesifer, F.R. and Cullen, W.H., *Fractographs Furnish Experimental Support for Cleavage Fracture Model*, Metallurgical Transactions A, Vol. 7A, pp. 1803–1806 (1976).

20. Curry, D.A., *Cleavage Micromechanisms of Crack Extension in Steels*, Metal Science, Vol. 14, pp. 319–326 (1980).

21. Yu, C. and Yan, M., *A Calculation of the Threshold Stress Intensity Range for Fatigue Crack Propagation in Metals*, Fatigue of Engineering Materials and Structures, Vol. 3, pp. 189–192 (1980).

22. Yoder, G.R., Cooley, L.A. and Crooker, T.W., *A Micromechanistic Interpretation of Cyclic Crack-Growth Behaviour in a Beta-Annealed Ti-6Al-4V Alloy*, US Naval Research Laboratory Report 8048, November 1976.

23. Sadananda, K. and Shahinian, P., *Prediction of Threshold Stress Intensity for Fatigue Crack Growth Using a Dislocation Model*, International Journal of Fracture, Vol. 13, pp. 585–594 (1977).

24. Taylor, D., *A Model for the Estimation of Fatigue Threshold Stress Intensities in Materials with Various Different Microstructures*, Fatigue Thresholds Fundamental and Engineering Applications, Engineering Materials Advisory Services, pp. 455–470 (1982): Warley, West Midlands, U.K.

25. Yoder, G.R., Cooley, L.A. and Crooker, T.W., *On Microstructural Control of Near-Threshold Fatigue Crack Growth in 7000-Series Aluminium Alloys*, US Naval Research Laboratory Memorandum Report 4787, April 1982.

26. Suresh, S., Parks, D.M., and Ritchie, R.O., *Crack Tip Oxide Formation and its Influence on Fatigue Thresholds*, Fatigue Thresholds Fundamental and Engineering Applications, Engineering Materials Advisory Services, pp. 391–408 (1982): Warley, West Midlands, U.K.

27. Yoder, G.R., Cooley, L.A. and Crooker, T.W., *Observations on the Generality of the Grain Size Effect on Fatigue Crack Growth in Alpha Plus Beta Titanium Alloys*, US Naval Research Laboratory Memorandum Report 4232, May 1980.

28. Yoder, G.R., Cooley, L.A. and Crooker, T.W., *A Critical Analysis of Grain-Size and Yield-Strength Dependence of Near-Threshold Fatigue-Crack Growth in Steels*, US Naval Research Laboratory Memorandum Report 4576, July 1981.

29. Stofanak, R.J., Hertzberg, R.W., Leupp, J. and Jaccard, R., *On the Cyclic Behaviour of Cast and Extruded Aluminium Alloys. Part B: Fractography*, Engineering Fracture Mechanics, Vol. 17, pp. 541–554 (1983).

30. Pelloux, R.M.N., *Crack Extension by Alternating Shear*, Engineering Fracture Mechanics, Vol. 1, pp. 697–704 (1970).

31. Kuo, A.S. and Liu, H.W., *An Analysis of Unzipping Model for Fatigue Crack Growth*, Scripta Metallurgica, Vol. 10, pp. 723–728 (1976).

32. Antolovich, S.D., Saxena, A. and Chanani, G.R., *A Model for Fatigue Crack Propagation*, Engineering Fracture Mechanics, Vol. 7, pp. 649–652 (1975).

33. Saxena, A. and Antolovich, S.D., *Low Cycle Fatigue, Fatigue Crack Propagation and Substructures in a Series of Polycrystalline Cu-Al Alloys*, Metallurgical Transactions A, Vol. 6A, pp. 1809–1828 (1975).

34. Green, J.A.S., Hayden, H.W. and Montague, W.G., *The Influence of Loading Mode on the Stress Corrosion Susceptibility of Various Alloy-Environment Systems*, Effect of Hydrogen on Behaviour of Materials, Metallurgical Society of AIME, pp 200–217 (1976): New York.

35. Wanhill, R.J.H., *Application of Fracture Mechanics to Cleavage in Highly Textured Titanium Alloys*, Fracture Mechanics and Technology, Sijthoff and Noordhoff, Vol.

1, pp. 563–572 (1977): Alphen aan den Rijn, The Netherlands.

36. Van Leeuwen, H.P., *Quantitative Models of Hydrogen-Induced Cracking in High-Strength Steel*, Reviews on Coatings and Corrosion, Vol. IV, pp. 5–93 (1979).

37. Van Leeuwen, H.P., *A Failure Criterion for Internal Hydrogen Embrittlement*, Engineering Fracture Mechanics, Vol. 9, pp. 291–296 (1977).

38. Beck, T.R., Blackburn, M.J. and Speidel, M.O., *Stress Corrosion Cracking of Titanium Alloys: SCC of Aluminium Alloys, Polarization of Titanium Alloys in HCl and Correlation of Titanium and Aluminium SCC Behaviour*, Quarterly Progress Report No. 11 Contract NAS 7-489, Boeing Scientific Research Laboratories (1969): Seattle, Washington.

39. Van Leeuwen, H.P., *Plateau Velocity of SCC in High Strength Steel − A Quantitative Treatment*, Corrosion-NACE, Vol. 31, pp. 42–50 (1975).

40. Smith, T., *A Capillary Model for Stress-Corrosion Cracking of Metals in Fluid Media*, Corrosion Science, Vol. 12, pp. 45–56 (1972).

41. Wei, R.P. and Landes, J.D., *Correlation between Sustained-Load and Fatigue Crack Growth in High-Strength Steels*, Materials Research and Standards, Vol. 9, pp. 25–27, 44, 46 (1969).

42. Austen, I.M. and Walker, E.F., *Quantitative Understanding of the Effects of Mechanical and Environmental Variables on Corrosion Fatigue Crack Growth Behaviour*, The Influence of Environment on Fatigue, Institution of Mechanical Engineers, pp. 1–10 (1977): London.

Index